
APPLIED ELECTRICITY AND ELECTRONICS DIVISION

General Editor: P. HAMMOND

Symmetrical Components

Symmetrical Components

BY

L. J. MYATT, B.SC. (ENG.), A.R.T.C.S., C.ENG., M.I.E.E.
Head of Department of Engineering and Building,
Thurso Technical College

1966
THE QUEEN'S AWARD TO INDUSTRY 1966

PERGAMON PRESS

OXFORD · LONDON · EDINBURGH · NEW YORK
TORONTO · SYDNEY · PARIS · BRAUNSCHWEIG

Pergamon Press Ltd., Headington Hill Hall, Oxford
4 & 5 Fitzroy Square, London W.1
Pergamon Press (Scotland) Ltd., 2 & 3 Teviot Place, Edinburgh 1
Pergamon Press Inc., 44–01 21st Street, Long Island City, New York 11101
Pergamon of Canada Ltd., 207 Queen's Quay West, Toronto 1
Pergamon Press (Aust.) Pty. Ltd., 19a Boundary Street,
Rushcutters Bay, N.S.W. 2011
Pergamon Press S.A.R.L., 24 rue des Écoles, Paris 5e
Vieweg & Sohn GmbH, Burgplatz 1, Braunschweig

First edition 1968
Library of Congress Catalog Card No. 68–26942

Printed in Great Britain by A. Wheaton & Co., Exeter

08 012978 1 (flexicover)
08 012979 X (hard cover

Contents

Preface

THE purpose of this book is to provide a text which introduces the student to the method of symmetrical components and to demonstrate the method by which an analysis may be made of simple power networks subjected to unbalanced fault conditions. It is written having in mind the particular needs of students preparing for final examinations of University Degree courses, the Higher National Certificate, the Council of Engineering Institutions, and other professional bodies. The subject-matter includes an evolution of the basic theory to an eventual solution of the conditions for unbalance on three-phase machines. This leads on to show the resolution of three-phase power networks into separate positive-, negative-, and zero-sequence networks and the use of these networks in order to solve for conditions of unbalanced faults. Chapters are also included to deal with various types of filter networks and the solution of problems involving networks having unbalanced impedances and similar associated topics.

As a result, it is expected that the text will also prove a useful reference for the practising engineer, whilst taking the student from the realms of theory into practice.

Acknowledgements

THE author wishes to acknowledge the assistance and help given in the preparation of this book by numerous people, including:

Dr. J. R. Mortlock for helpful guidance and valuable suggestions.

The Institution of Electrical Engineers for permission to use past examination questions. The Institution is, of course, in no way responsible for either the accuracy of the answers or the method of solution.

Mrs. D. Harrison and Mrs. H. Powell, who typed the manuscript, and also my wife, who assisted in checking it.

Thurso 1968 L.J.M.

CHAPTER 1

The h Operator

In a three-phase system, the three-phase voltages or currents may be said to be balanced if they are sinusoidal and, when represented by vectors, are of equal magnitude and displaced from each other by equal phase angles of 120°. A three-phase circuit is also balanced, or symmetrical, when each of the three phases contain equal impedances. Furthermore, if a balanced system of three-phase voltages is applied across the terminals of a balanced three-phase network, then the currents flowing in each of the three phases will also be balanced.

The solution of a problem involving a balanced three-phase system may be found by considering only one phase in which the voltage, or current, is taken as the reference vector and solving for that phase alone as in the case of a single-phase problem. The magnitude of the currents and voltages in the other two phases will then be the same as in the reference phase but there will be a corresponding phase displacement of $\pm 120°$ as appropriate.

If now the applied voltage is unbalanced, or the impedances in each phase are no longer identical, the three-phase currents will also become unbalanced and the problem cannot be solved by considering a single phase as in the above method. It would be possible in a simple problem of unbalance to find a solution by applying Kirchhoff's laws, but in the more involved type of problem, such as that involving a power network, a less laborious method is necessary.

Considerable thought was given to the problem of unbalance, at the beginning of the century, with particular reference to the unbalanced loading of three-phase machines, and L. G. Stokvis showed that, under such conditions, the armature m.m.f. of a three-phase generator may be treated as two separate components, equivalent to the effect of two balanced loads. It remained for C. L. Fortesque to develop the method, now known as the method of symmetrical components, after working on problems associated with the use of phase balancers in the single-phase electrification of railways. Following on from this, he studied the problem of unbalance in general when, in 1918, he published his very important paper "The method of symmetrical coordinates applied to the solution of polyphase networks". In this paper he showed how it was possible for a set of unbalanced polyphase currents to be resolved into a number of component systems of balanced currents equal to the number of phases. A further development of the theory provides considerable simplification in solving problems involving unbalanced faults on power networks.

1. Representation of sinusoidal quantity by a vector

The instantaneous value v of sinusoidal voltage of frequency f Hz is given by

$$v = V \sin \omega t, \tag{1}$$

where V = maximum value and $\omega = 2\pi f$.

This may be represented by a rotating vector of magnitude V having a constant angular velocity ω rad/s as shown in Fig. 1. The vector OV rotates anticlockwise about O starting from Ox when $t = 0$. The angle turned through, at any time t, is then given by ωt and the instantaneous value v at any time is given by $V \sin \omega t$ which is the projection of OV on the Y-axis, Oy.

If the vector V is now shown on an Argand diagram as in Fig. 2 where $\theta = \omega t$, it may be represented by $(a + jb)$.

where
$$a = V \cos \theta,$$
$$b = V \sin \theta,$$
$$a + jb = V(\cos \theta + j \sin \theta). \tag{2}$$

Another way of expressing the vector is as $V\underline{/\theta}$, where V is known as the *modulus* and would correspond to the peak value of the alternating voltage wave of Fig. 1, and θ is known as the *argument* and represents the angular displacement of the vector V from the real axis $X'OX$.

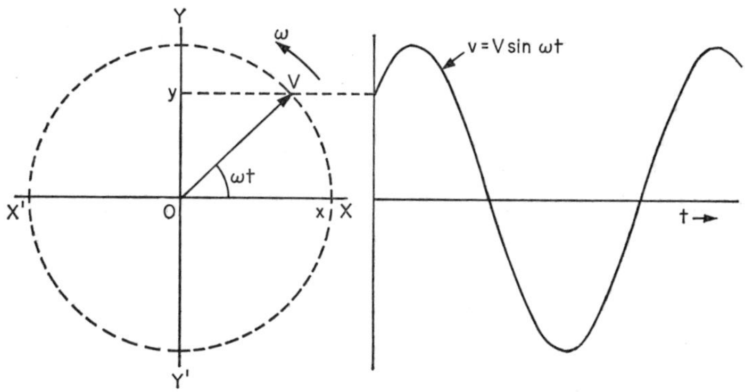

FIG. 1. Vector representation of a sinusoidal quantity.

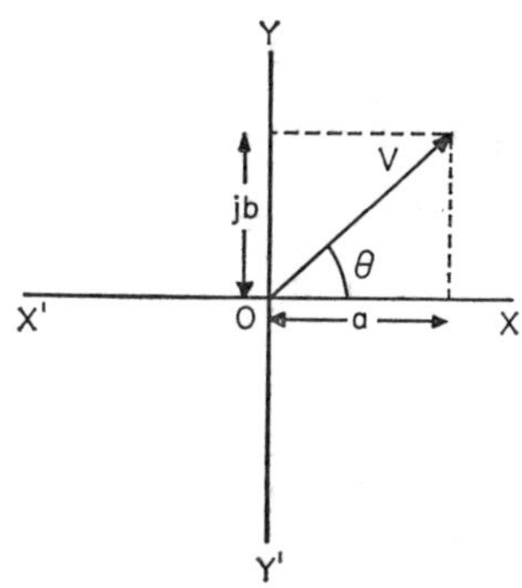

FIG. 2. Representation of a vector on an Argand diagram.

Now
$$\cos\theta + j\sin\theta = e^{j\theta}, \tag{3}$$
and
$$V(\cos\theta + j\sin\theta) = Ve^{j\theta}. \tag{4}$$
Thus the expression $Ve^{j\theta}$ may be used to represent a vector of modulus V and argument θ.

2. The *h* operator

When performing calculations on single-phase circuits use was made of an operator, which, when multiplied by a vector, had the effect of rotating the vector through an angle of 90°, and for this purpose the *j* operator was used. In dealing with three-phase quantities, and in particular when applying the method of symmetrical components, there are definite advantages to be gained by having an operator which is capable of rotating a vector through an angle of 120°, due to the angular displacement of balanced three-phase vectors.

First let us consider the multiplication of two complex quantities which may be represented by $r_1 \underline{/\theta_1}$ and $r_2 \underline{/\theta_2}$, then we may write

$$\begin{aligned} r_1 \underline{/\theta_1}\, r_2 \underline{/\theta_2} &= r_1 (\cos\theta_1 + j\sin\theta_1)\, r_2 (\cos\theta_2 + j\sin\theta_2) \\ &= r_1 r_2 (\cos\theta_1 \cos\theta_2 + j\sin\theta_1 \cos\theta_2 \\ &\quad + j\cos\theta_1 \sin\theta_2 - \sin\theta_1 \sin\theta_2) \\ &= r_1 r_2 (\cos\theta_1 \cos\theta_2 - \sin\theta_1 \sin\theta_2) \\ &\quad + j(\sin\theta_1 \cos\theta_2 + \cos\theta_1 \sin\theta_2) \\ &= r_1 r_2 [\cos(\theta_1 + \theta_2) + j\sin(\theta_1 + \theta_2)] \\ &= r_1 r_2 \underline{/\theta_1 + \theta_2}. \end{aligned} \tag{5}$$

Thus it may be observed that the modulus of the product of two complex quantities is equal to the product of their moduli, and that the argument of their product is equal to the sum of their arguments.

Suppose now that $r_2 \underline{/\theta_2}$ is replaced by a voltage vector $V\underline{/\theta}$, and $r_1 \underline{/\theta_1}$ is used to denote an operator such that it will rotate the voltage vector through an angle of 120° to give a new vector $V \underline{/\theta + 120°}$, then

$$r_1 \underline{/\theta_1}\, V\underline{/\theta} = V\underline{/\theta + 120°}.$$

Comparing the moduli and arguments with the results of eqn. (5) gives

$$r_1 V = V.$$

Therefore $$r_1 = 1;$$

also $$\theta + \theta_1 = \theta + 120°$$

Therefore $$\theta_1 = 120°.$$

The operator $r_1 \underline{/\theta_1}$ is then given by

$$r_1 \underline{/\theta_1} = 1\underline{/120°} = e^{j\,2\pi/3}. \tag{6}$$

Various symbols, including a, λ, and h, have been assigned to this operator by different writers on the subject, but the latter is now used in accordance with standard practice, as recommended by the British Standards Institution. Hence,

$$\begin{aligned} h &= e^{j\,2\pi/3}, \\ &= \cos 2\pi/3 + j \sin 2\pi/3, \\ &= -\tfrac{1}{2} + j\tfrac{1}{2}\sqrt{3}. \end{aligned} \tag{7}$$

Considering again the results of eqns. (5) and (6) it may be stated that, in general, a vector multiplied by $e^{j\pi/n}$ will cause an angular rotation of $2\pi/n$ in the vector, and so it is a simple matter to produce an operator capable of rotating a vector through any desired angle. For example, the j operator which rotates a vector through an angle of 90° would be represented by $e^{j\pi/2}$, where

$$\begin{aligned} e^{j\pi/2} &= (\cos \pi/2 + j \sin \pi/2) \\ &= j. \end{aligned} \tag{8}$$

A further significance may be attached to the h operator if we consider a vector represented by V as shown in Fig. 3 and which is shown drawn on the reference axis OX. If this vector is multiplied by h, giving hV, it will be rotated through 120°. A second multiplication by h, giving h^2V, will give a further rotation of 120° and a third multiplication by h, giving h^3V, will cause a further rotation of 120° bringing us back to the original position of vector V on the reference axis. It is now apparent that

$$h^3 V = V,$$
$$h^3 = 1. \tag{9}$$

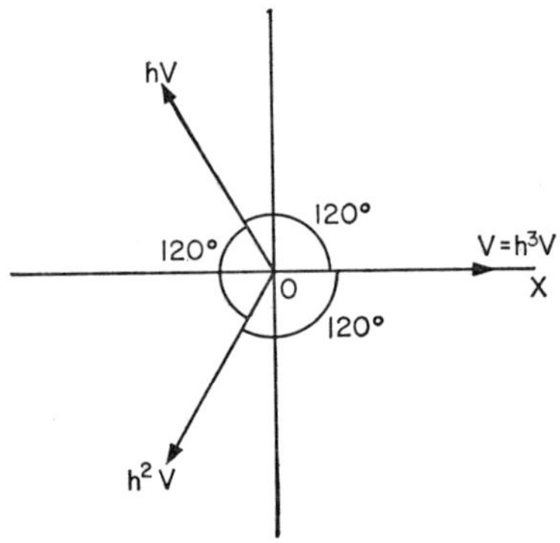

FIG. 3. The h operator.

In other words, h must be represented by one of the three cube roots of unity, which may be found in the following manner.

Let $$x = \sqrt[3]{1},$$
then $$x^3 = 1,$$
$$x^3 - 1 = 0,$$
$$(x - 1)(x^2 + x + 1) = 0.$$

One solution is given by $x = 1$, and the other two roots are given by solving the quadratic equation

$$x^2 + x + 1 = 0,$$

i.e. $$x = \frac{-1 \pm \sqrt{(1-4)}}{2}$$
$$= -\tfrac{1}{2} \pm j\tfrac{1}{2}\sqrt{3}$$

Thus the three cube roots of unity are given by

$$1, \quad (-\tfrac{1}{2} + j\tfrac{1}{2}\sqrt{3}), \quad (-\tfrac{1}{2} - j\tfrac{1}{2}\sqrt{3}), \tag{10}$$

and from eqn. (7) it is seen that it is the second of the three roots which has the particular value of h.

3. Functions of h

Now we have shown that

$$h = -\tfrac{1}{2} + j\tfrac{1}{2}\sqrt{3}.$$

Then
$$h^2 = (-\tfrac{1}{2} + j\tfrac{1}{2}\sqrt{3})^2,$$
$$= \tfrac{1}{4} - j\tfrac{1}{2}\sqrt{(3)} - \tfrac{3}{4},$$
$$= -\tfrac{1}{2} - j\tfrac{1}{2}\sqrt{3}; \tag{11}$$

also

$$h^3 = h\,h^2,$$
$$= (-\tfrac{1}{2} + j\tfrac{1}{2}\sqrt{3})(-\tfrac{1}{2} - j\tfrac{1}{2}\sqrt{3}),$$
$$= \tfrac{1}{4} + j\tfrac{1}{4}\sqrt{(3)} - j\tfrac{1}{4}\sqrt{(3)} + \tfrac{3}{4},$$
$$= 1, \tag{12}$$

which is the result given in eqn. (9).

From eqns. (7) and (11) we have

$$h - h^2 = (-\tfrac{1}{2} + j\tfrac{1}{2}\sqrt{3}) - (-\tfrac{1}{2} - j\tfrac{1}{2}\sqrt{3}),$$
$$= j\sqrt{3}. \tag{13}$$

Also from eqns. (7) and (11) we have

$$1 + h + h^2 = 1 + (-\tfrac{1}{2} + j\tfrac{1}{2}\sqrt{3}) + (-\tfrac{1}{2} - j\tfrac{1}{2}\sqrt{3}),$$
$$= 0. \tag{14}$$

This result is to be expected since 1, h, and h^2 represent three vectors each of unit length with equal phase angles of 120° between them, i.e. three symmetrical vectors.

From eqn. (14) it also follows that

$$1 + h = -h^2, \tag{15}$$

$$1 + h^2 = -h. \tag{16}$$

Various functions of h are shown represented on the Argand diagram of Fig. 4.

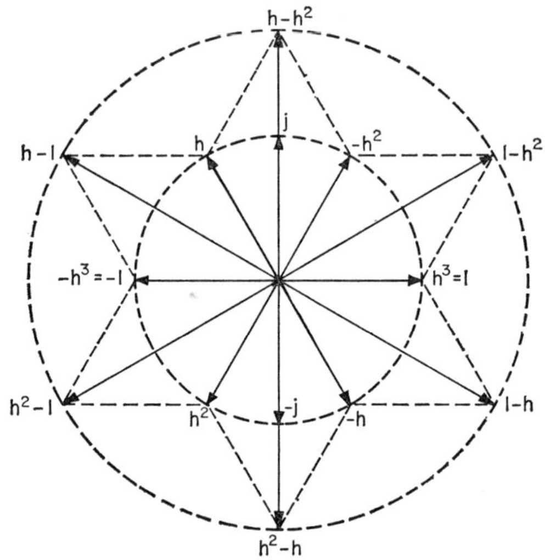

FIG. 4. Functions of h.

4. Simplification of h^n

In simplifying an expression containing h^n the following procedure may be adopted.

If n is a positive integer, and since $h^3 = 1$, put

$$h^n = \frac{h^n}{(h^3)^m},$$

where m is a positive integer such that the solution yields 1, h or h^2.

EXAMPLE 1. To simplify h^7.

We put $m = 2$ such that

$$h^7 = \frac{h^7}{(h^3)^2} = h. \tag{17}$$

If n is a negative integer we may write

$$h^n = h^n\,(h^3)^m,$$

where m again is a positive integer.

EXAMPLE 2. To simplify h^{-1}.

We put $m = 1$ such that

$$h^{-1} = h^{-1}\,h^3 = h^2. \qquad (18)$$

Where n is not an integer it is usually better to revert to the polar form of h when

$$h = 1\,\underline{/120^\circ}.$$

Then $$h^n = 1\,\underline{/n \times 120^\circ}.$$

EXAMPLE 3. To simplify $h^{\frac{3}{4}}$.

We write
$$\begin{aligned} h^{\frac{3}{4}} &= 1\,\underline{/\tfrac{3}{4} \times 120^\circ}, \\ &= \cos 90^\circ + j \sin 90^\circ, \\ &= j. \end{aligned} \qquad (19)$$

EXAMPLE 4. To simplify $h^{-\frac{1}{2}}$.

We write
$$\begin{aligned} h^{-\frac{1}{2}} &= 1\,\underline{/-\tfrac{1}{2} \times 120^\circ}, \\ &= 1\,\underline{/-60^\circ}, \\ &= \cos(-60^\circ) + j \sin(-60^\circ), \\ &= \tfrac{1}{2} - j\tfrac{1}{2}\sqrt{3}, \\ &= -h. \end{aligned} \qquad (20)$$

5. Simplification of $(h + k)^{-1}$ and $(h^2 + k)^{-1}$

Expressions of the above form, in which k is a real number, could be simplified by rewriting the expression in terms of the

j operator and then multiplying the numerator and denominator by the conjugate of the denominator in order to remove all the j terms from the denominator. It is, however, often preferable to work in terms of the h operator alone as the whole purpose of using such an operator is to simplify the arithmetic.

Simplification of terms containing a function of h in the denominator may be carried out by a method similar to that used for the j operator. Considering the expression $(h + k)^{-1}$ it is possible to remove the h term from the denominator by multiplying the numerator and denominator by $(h - k + 1)$. Hence,

$$\frac{1}{h+k} = \frac{h-k+1}{(h+k)(h-k+1)},$$

$$= \frac{h-k+1}{h^2 - kh + h + kh - k^2 + k},$$

$$= \frac{h-k+1}{h^2 + h - k^2 + k},$$

and, using the relationship $(1 + h + h^2) = 0$, this gives

$$\frac{1}{h+k} = \frac{h-k+1}{k-k^2-1}. \tag{21}$$

EXAMPLE 5. To simplify $(h + 2)^{-1}$.

We write

$$\frac{1}{h+2} = \frac{h-1}{(h+2)(h-1)},$$

$$= \frac{h-1}{h^2 - h + 2h - 2},$$

$$= \frac{h-1}{h^2 + h - 2},$$

$$= -\tfrac{1}{3}(h-1). \tag{22}$$

Similarly, an expression of the form $(h^2 + k)^{-1}$ may be simplified by multiplying the numerator and denominator by $(h^2 - k + 1)$ as follows:

$$\begin{aligned}\frac{1}{h^2 + k} &= \frac{h^2 - k + 1}{(h^2 + k)(h^2 - k + 1)},\\ &= \frac{h^2 - k + 1}{h^4 - kh^2 + h^2 + kh^2 - k^2 + k},\\ &= \frac{h^2 - k + 1}{h^2 + h - k^2 + k},\\ &= \frac{h^2 - k + 1}{k - k^2 - 1}. \qquad (23)\end{aligned}$$

EXAMPLE 6. To simplify $(h^2 + 2)^{-1}$.

We write
$$\begin{aligned}\frac{1}{h^2 + 2} &= \frac{h^2 - 1}{(h^2 + 2)(h^2 - 1)},\\ &= \frac{h^2 - 1}{h^4 - h^2 + 2h^2 - 2},\\ &= \frac{h^2 - 1}{h^2 + h - 2},\\ &= -\tfrac{1}{3}(h^2 - 1). \qquad (24)\end{aligned}$$

6. Useful functions of h

In Table 1 are given a number of the more useful functions of the h operator, some of which have already been derived.

TABLE 1. USEFUL FUNCTIONS OF h.

Function	Exponential form	Polar form	Symbolic form
h	$e^{j2\pi/3}$	$1/\underline{120^\circ}$	$-\frac{1}{2}+j\frac{1}{2}\sqrt{3}$
h^2	$e^{j4\pi/3}$	$1/\underline{240^\circ}$	$-\frac{1}{2}-j\frac{1}{2}\sqrt{3}$
$h^3 = 1$	$e^{j2\pi} = ej^{0}$	$1/\underline{360^\circ} = 1/\underline{0^\circ}$	$1+j0$
$h^4 = h$	$e^{j8\pi/3} = e^{j2\pi/3}$	$1/\underline{480^\circ} = 1/\underline{120^\circ}$	$-\frac{1}{2}+j\frac{1}{2}\sqrt{3}$
$h^{-1} = h^2$	$e^{-j2\pi/3} = e^{j4\pi/3}$	$1/\underline{-120^\circ} = 1/\underline{240^\circ}$	$-\frac{1}{2}-j\frac{1}{2}\sqrt{3}$
$h^{-2} = h$	$e^{-j4\pi/3} = e^{j2\pi/3}$	$1/\underline{-240^\circ} = 1/\underline{120^\circ}$	$-\frac{1}{2}+j\frac{1}{2}\sqrt{3}$
$h^{-3} = 1$	$e^{-j2\pi} = e^{j0}$	$1/\underline{-360^\circ} = 1/\underline{0^\circ}$	$1+j0$
$1 - h^2$	$\sqrt{(3)}e^{j\pi/6}$	$\sqrt{3}/\underline{30^\circ}$	$\frac{3}{2}+j\frac{1}{2}\sqrt{3}$
$h - h^2$	$\sqrt{(3)}e^{j\pi/2}$	$\sqrt{3}/\underline{90^\circ}$	$0+j\sqrt{3}$
$h - 1$	$\sqrt{(3)}e^{j5\pi/6}$	$\sqrt{3}1/\underline{50^\circ}$	$-\frac{3}{2}+j\frac{1}{2}\sqrt{3}$
$1 + h = -h^2$	$-e^{j4\pi/3} = e^{j\pi/3}$	$-1/\underline{240^\circ} = 1/\underline{60^\circ}$	$\frac{1}{2}+j\frac{1}{2}\sqrt{3}$
$1 + h^2 = -h$	$-e^{j2\pi/3} = e^{j5\pi/3}$	$-1/\underline{120^\circ} = 1/\underline{300^\circ}$	$\frac{1}{2}-j\frac{1}{2}\sqrt{3}$
$h + h^2 = -1$	$-e^{j0} = e^{j\pi}$	$-1/\underline{0^\circ} = 1/\underline{180^\circ}$	$-1+j0$
$(1 + h)^{-1} = -h$	$-e^{j2\pi/3} = e^{j5\pi/3}$	$-1/\underline{120^\circ} = 1/\underline{300^\circ}$	$\frac{1}{2}-j\frac{1}{2}\sqrt{3}$
$(1 + h^2)^{-1} = -h^2$	$-e^{j4\pi/3} = e^{j\pi/3}$	$-1/\underline{240^\circ} = 1/\underline{60^\circ}$	$\frac{1}{2}+j\frac{1}{2}\sqrt{3}$
$1 + h + h^2$			$0+j0$

7. Symmetrical three-phase voltages

In the discussion which follows, what is said of three-phase voltages applies also to three-phase currents and may be interpreted as such by reading current for voltage.

Any set of three-phase voltages, which are sinusoidal, are said to be symmetrical if their maximum values are equal, and there is the same phase angle of $2\pi/3$, $4\pi/3$, or 2π between each voltage and the next. This phase angle, sometimes referred to as the characteristic phase angle, will give rise to a different set of three-phase voltage vectors depending upon its value.

8. Positive-phase sequence

In a three-phase system, in which the three-phase quantities are normally referred to as red, yellow, and blue, for balanced conditions, the three-phase symmetrical voltages would normally reach their maximum positive values in the sequence red, yellow, blue, red as shown in Fig. 5. The characteristic phase angle in this case is seen to be $2\pi/3$ radians or 120°, and the voltages are said to have positive-phase sequence. This is commonly referred to as the standard phase sequence.

In applying the method of symmetrical components it is necessary to identify the different phase sequence components, and for this reason it is usual to use the subscript 1 to denote positive

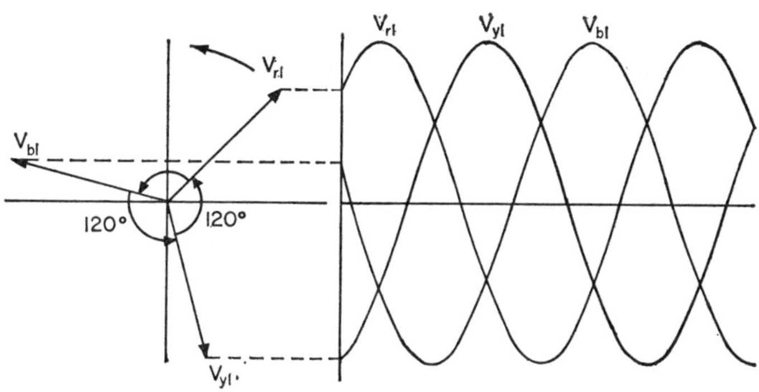

FIG. 5. Positive-phase sequence.

sequence components. Thus the three-phase voltage vectors having positive phase sequence would be referred to as V_{r1}, V_{y1}, and V_{b1}, and, because of their vector relationship, may be written as V_{r1}, h^2V_{r1}, and hV_{r1}.

9. Negative phase sequence

If the characteristic phase angle is now made equal to $4\pi/3$ rad., or 240°, a new set of vectors will be defined as shown in Fig. 6, where the voltages may now be seen to reach their maximum values in the order red, blue, yellow, red. Such a set of three-phase voltages is said to have negative-phase sequence.

It is important to note here that the direction of rotation of the three voltage vectors in this case is again counter-clockwise, as in the case of the positive-phase sequence voltages. The only change made is in the angular displacement of the vectors and not the direction of rotation.

The subscript 2 is used to identify negative-phase sequence quantities so that the three-phase voltage vectors are referred to as V_{r2}, V_{y2}, and V_{b2} and may be written as V_{r2}, hV_{r2}, and h^2V_{r2}.

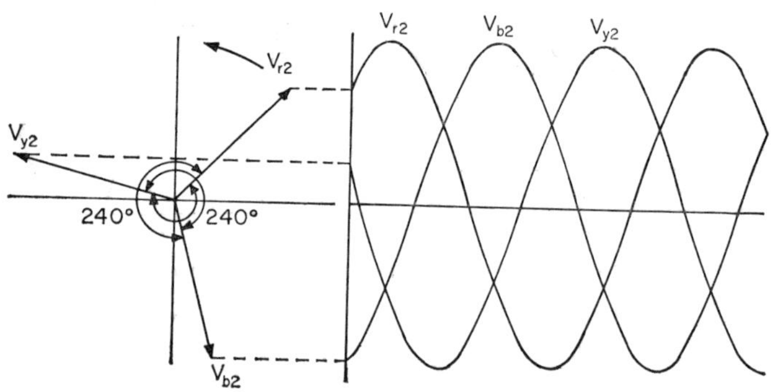

FIG. 6. Negative-phase sequence.

10. Zero-phase sequence

When the characteristic phase angle is made equal to 2π rad., or 360°, exactly the same effect is produced as if the phase angle were zero because the three-phase voltage vectors now appear to be coincident, i.e. they are in phase with each other and will each reach its maximum value at the same time. This is shown in Fig. 7.

Zero-phase sequence quantities are identified by the subscript 0 and the three-phase voltages would be written as V_{r0}, V_{y0}, V_{b0}, where $V_{r0} = V_{y0} = V_{b0}$.

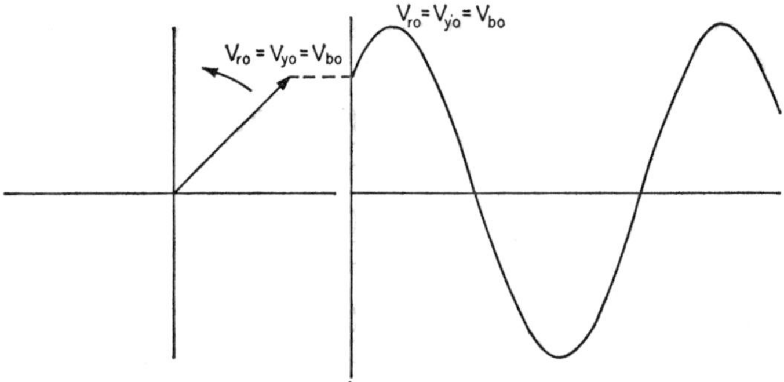

FIG. 7. Zero-phase sequence.

11. Star–delta impedance transformations

Any three-terminal network consisting of three impedances connected in delta may be replaced by an equivalent three-terminal network of three impedances connected in star, and vice versa. This property is often very useful in network reduction when resolving a complete network into a single equivalent impedance.

In Fig. 8 are shown three impedances Z_A, Z_B, and Z_C connected in delta and the three equivalent impedances Z_a, Z_b, and Z_c connected in star. For the two networks to be equivalent, the total

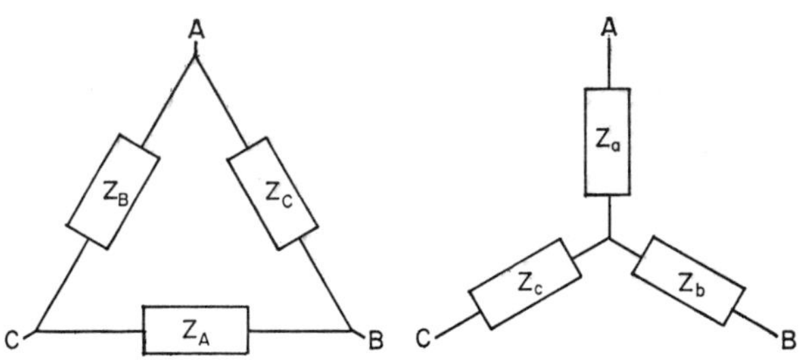

FIG. 8. Equivalent delta- and star-connected impedances.

impedance between A and B, B and C, and C and A, must be the same in both networks. Thus considering the two networks we have

$$Z_a + Z_b = \frac{Z_C(Z_A + Z_B)}{Z_A + Z_B + Z_C}, \tag{25}$$

$$Z_b + Z_c = \frac{Z_A(Z_B + Z_C)}{Z_A + Z_B + Z_C}, \tag{26}$$

$$Z_c + Z_a = \frac{Z_B(Z_C + Z_A)}{Z_A + Z_B + Z_C}. \tag{27}$$

Adding eqns. (25) and (27) and subtracting (26) gives

$$2\,Z_a = \frac{Z_C(Z_A + Z_B) + Z_B(Z_C + Z_A) - Z_A(Z_B + Z_C)}{Z_A + Z_B + Z_C},$$

$$Z_a = \frac{Z_B Z_C}{Z_A + Z_B + Z_C}. \tag{28}$$

Similarly,

$$Z_b = \frac{Z_C Z_A}{Z_A + Z_B + Z_C}, \tag{29}$$

$$Z_c = \frac{Z_A Z_B}{Z_A + Z_B + Z_C}. \tag{30}$$

Equations (28), (29), and (30) give the equivalent impedance values for the three star-connected impedances for conversion from delta to star which is the usual conversion to have to make. On the other hand, it may be necessary on occasions to make the conversion from star to delta as follows;

From eqns. (28) and (29): $\dfrac{Z_A}{Z_B} = \dfrac{Z_b}{Z_a}$. (31)

From eqns. (29) and (30): $\dfrac{Z_B}{Z_C} = \dfrac{Z_c}{Z_b}$. (32)

From eqns. (28) and (30): $$\frac{Z_C}{Z_A} = \frac{Z_a}{Z_c}. \tag{33}$$

Also from eqns. (29), $$Z_A = Z_b \left[\frac{Z_A}{Z_C} + \frac{Z_B}{Z_C} + 1\right],$$
$$= Z_b \left[\frac{Z_c}{Z_a} + \frac{Z_c}{Z_b} + 1\right],$$
$$= \frac{Z_b Z_c}{Z_a} + Z_c + Z_b.$$

Therefore $$Z_A = \frac{Z_a Z_b + Z_b Z_c + Z_c Z_a}{Z_a}. \tag{34}$$

Similarly,

$$Z_B = \frac{Z_a Z_b + Z_b Z_c + Z_c Z_a}{Z_b}, \tag{35}$$

$$Z_C = \frac{Z_a Z_b + Z_b Z_c + Z_c Z_a}{Z_c}. \tag{36}$$

12. Percentage and per unit values

Single-phase quantities

Calculations on a simple circuit, without transformers, containing a source of e.m.f. connected across an impedance are carried out using the units of volts, amperes, and ohms and are related by

$$V = IZ.$$

If now one or more transformers are included in the circuit, the calculation could be carried out by referring all the impedances and currents to a common base voltage and then performing the calculation as in the previous case. The resulting quantities thus obtained would then be referred back to the appropriate voltage. The voltage taken as the base value is purely arbitrary but in practice would probably be the voltage at some convenient place in the circuit.

A further method of calculation is by the use of percentage values, where all quantities are expressed as a percentage of some base value. For example, a voltage V may be expressed as a percentage value by taking some other voltage V_b as the base value, then

$$V\,(\text{in }\%) = \frac{V}{V_b} \times 100, \tag{37}$$

where V_b, of course, represents 100% voltage.

Similarly, a current I may be expressed as a percentage value in terms of a base current I_b, where

$$I\,(\text{in }\%) = \frac{I}{I_b} \times 100. \tag{38}$$

This being so, the base value of impedance is now defined such that, when base current flows through it, there will be a voltage drop across it of base voltage, i.e.

$$Z_b = \frac{V_b}{I_b}\text{ ohms}, \tag{39}$$

and any other value of impedance Z ohms will then be defined as a percentage value by

$$Z\,(\text{in }\%) = \frac{Z}{Z_b} \times 100,$$

$$= \frac{I_b Z}{V_b} \times 100. \tag{40}$$

Three base values of current, voltage, and impedance, have so far been used, the first two of which may now be used to define base kVA or base MVA, i.e.

$$MVA_b = V_b I_b \times 10^{-6}.$$

Hence we now have four base quantities, any two of which may be used to define the other two. In practice it is usual to take the

MVA (or *kVA*) and the voltage as the base quantities since equipment is normally rated in terms of *MVA* and in general the voltage variation is normally small for various parts of a system, and so voltage may be used as a useful reference quantity.

Taking now these quantities as base values we have

$$I_b = \frac{MVA_b \times 10^3}{kV_b} \text{ amp,} \tag{41}$$

$$Z_b = \frac{V_b}{I_b} = \frac{(kV_b)^2}{MVA_b} \text{ ohms,} \tag{42}$$

where all base quantities are phase values. Also,

$$Z \text{ (in \%)} = \frac{Z}{Z_b} \times 100,$$

$$= \frac{Z \times MVA_b}{(kV_b)^2} \times 100. \tag{43}$$

Base power in *MW* will be numerically equal to the base *MVA*.

Having defined our base quantities, supposing now that a current of $i\%$ flows through an impedance of $z\%$ then the voltage drop $v\%$ is given by

$$v \text{ (in \%)} \frac{i \times z}{100},$$

from which it is apparent that care must be taken to include the factor of 100 in the denominator.

Alternatively the current would be given by

$$i \text{ (in \%)} = \frac{v}{z} \times 100,$$

and the factor of 100 now appears in the numerator. In order to eliminate this factor of 100 it is more convenient to work in per unit values instead of percentage values where

$$\text{per unit value} = \frac{\text{percentage value}}{100}. \tag{44}$$

Hence we have

$$V_{p.u.} = \frac{V}{V_b}, \tag{45}$$

$$I_{p.u.} = \frac{I}{I_b},$$

$$= \frac{I \times kV_b}{MVA_b} \times 10^{-3}, \tag{46}$$

$$Z_{p.u.} = \frac{Z}{Z_b},$$

$$= \frac{I_b Z}{V_b},$$

$$= \frac{Z \times MVA_b}{(kV_b)^2}, \tag{47}$$

where V_b, I_b, and Z_b are the base values as before and

$$V_{p.u.} = I_{p.u.} \times Z_{p.u.} \tag{48}$$

One of the main advantages of the use of per unit, or percentage, values over ohmic values is that whereas the reactances of transformers—and machines—will have a very wide variation over the range of ratings, when expressed in ohmic values the variation is comparatively small when expressed in per unit or percentage values based upon the rated *MVA* and voltage of the equipment. This also gives a basis of comparison of equipment of different ratings, and at the same time if the reactance of some particular piece of equipment in a network is unknown it allows an assumed value to be used without any great loss of accuracy.

Three-phase quantities

Whereas with single-phase equipment the base voltage and *MVA* are phase values, in the case of three-phase equipment the rated voltage is usually the line-to-line voltage and rated MVA the three-phase value. Taking these now as the base values gives

$$Z_b = \frac{(kV_b)^2}{MVA_b} \text{ ohms} \tag{49}$$

as in the single-phase case, but now the base current is given by

$$I_b = \frac{MVA_b \times 10^3}{\sqrt{3}\, kV_b} \text{ amp} \tag{50}$$

and also

$$Z_{p.u.} = \frac{Z \times MVA_b}{(kV_b)^2}, \tag{51}$$

$$I_{p.u.} = \frac{\sqrt{3} I \times kV_b}{MVA_b} \times 10^{-3}. \tag{52}$$

In Table 2 are shown the relationships between the single-phase and three-phase base values

TABLE 2. PER UNIT BASE VALUES.

Base quantity	Base value	
	Single-phase	Three-phase
MVA	Single-phase value	Three-phase value
Voltage	Line-to-neutral value	Line-to-line value
Current	$\frac{MVA_b \times 10^3}{kV_b}$	$\frac{MVA_b \times 10^3}{\sqrt{3}\, kV_b}$
Impedance	$\frac{(kV_b)^2}{MVA_b}$	$\frac{(kV_b)^2}{MVA_b}$

Change of base values

In general the different items of plant in a network will have different MVA ratings and before making calculations it is necessary to convert all per unit impedances to the same MVA base.

The base value of *MVA* chosen is purely arbitrary and could be that of the largest machine, or any other convenient value such as 10, 50, or 100 MVA.

If $Z_{p.u.1}$ is the per unit value of an impedance on an *MVA* base of MVA_{b1} then from eqn. (49) it may be seen that the per unit value of impedance $Z_{p.u.2}$ on a new *MVA* base MVA_{b2} is given by

$$Z_{p.u.2} = Z_{p.u.1} \frac{MVA_{b2}}{MVA_{b1}}. \tag{53}$$

Similarly, per unit impedances on a given voltage base may be expressed on a new voltage base by

$$Z_{p.u.2} = Z_{p.u.1} \frac{(kV_{b1})^2}{(kV_{b2})^2}. \tag{54}$$

The following basic rules apply when using per unit quantities:

(1) All impedances must be on the same *MVA* base.

(2) In a network not containing transformers the base voltage will be the same throughout the network

(3) If the network contains transformers the base voltage must be changed in proportion to the turns ratio of the transformer when passing from the primary winding to the secondary winding and vice versa. Thus in a network containing a single transformer, rated at 50 MVA 13,800/132,000 V, the base voltage for the network on the primary side, and including the transformer primary leakage impedance, would be 13·8 kV, and for the network and transformer leakage impedance on the secondary side the base voltage would be 132 kV.

Supposing now that in the above transformer

$$\text{Primary leakage reactance} = 0{\cdot}2\ \Omega.$$

$$\text{Secondary leakage reactance} = 18\ \Omega.$$

Then, applying eqn. (51),

$$\text{Primary leakage reactance p.u.} = \frac{0 \cdot 2 \times 50}{13 \cdot 8^2} = 0 \cdot 0525$$

$$\text{Secondary leakage reactance p.u.} = \frac{18 \times 50}{132^2} = 0 \cdot 0516$$

$$\text{Total leakage reactance p.u.} = \overline{\underline{0 \cdot 1041}}$$

Now in the above calculation the appropriate base voltage has been used for both the primary and secondary winding. The same result would ensue if all the reactance of both windings had been referred to either the primary or secondary. For example,

Secondary reactance referred to primary

$$= 18 \left(\frac{13 \cdot 8}{132}\right)^2 = 0 \cdot 197 \ \Omega.$$

Therefore total leakage reactance referred to primary

$$= 0 \cdot 2 + 0 \cdot 197 = 0 \cdot 397 \ \Omega.$$

$$\text{Therefore total leakage reactance p.u.} = \frac{0.397 \times 50}{13 \cdot 8^2} = 0 \cdot 1041.$$

Hence the same result is achieved as by the previous method, and, further, the same result would be obtained by referring the total reactance to the secondary winding.

Exercises

1. Simplify the following and express in symbolic form: $(1 + h)(2 + h^2)$,

$$(1 - h)^{-1}, \quad \frac{1 - h^2}{1 + 2h}, \quad \text{where } h = (-\tfrac{1}{2} + j\tfrac{1}{2}\sqrt{3}).$$

2. Three impedances, $Z_A = (1 + j3)$ ohms, $Z_B = (2 + j5)$ ohms, and $Z_C = (5 - j2)$ ohms are connected in delta forming a three-terminal network. Calculate the values of three star-connected impedances which will form an equivalent network.

3. A generator reactance when expressed on a 75 MVA base is given as $0 \cdot 15$ per unit. What would be the per unit values of reactance when expressed on 100 MVA and 50 MVA bases?

CHAPTER 2

Symmetrical Components

FORTESQUE, in his paper referred to in Chapter 1, showed that "a system of n vectors or quantities may be resolved when n is prime into n different symmetrical groups or systems, one of which consists of n equal vectors and the remaining $(n - 1)$ systems consist of n equi-spaced vectors which with the first mentioned groups of equal vectors forms an equal number of symmetrical n-phase systems. When n is not prime some of the n-phase systems degenerate into repetitions of systems having numbers of phases corresponding to the factors of n." Thus he produced a method of resolving an unbalanced system of polyphase vectors into a number of balanced systems which forms the basis of the method of symmetrical components.

In terms of a three-phase system this means that an unbalanced set of voltages or currents can be resolved into two symmetrical three-phase systems having opposite phase sequence (i.e. positive- and negative-phase sequence) plus a third set of equal vectors having zero-phase displacement (i.e. zero-phase sequence).

Although the method of symmetrical components may be applied to any n-phase system, only three-phase systems are to be considered here as this is the more usual case.

14. Resolution into symmetrical components

Let V_a, V_b, and V_c be three unbalanced three-phase voltages, and since $1 + h + h^2 = 0$, the following equations may be proved by inspection:

$$\left.\begin{aligned}
V_a &= \\
&\frac{V_a + V_b + V_c}{3} + \frac{V_a + hV_b + h^2V_c}{3} + \frac{V_a + h^2V_b + hV_c}{3}, \\
V_b &= \\
&\frac{V_a + V_b + V_c}{3} + h^2 \frac{V_a + hV_b + h^2V_c}{3} + h \frac{V_a + h^2V_b + hV_c}{3}, \\
V_c &= \\
&\frac{V_a + V_b + V_c}{3} + h \frac{V_a + hV_b + h^2V_c}{3} + h^2 \frac{V_a + h^2V_b + hV_c}{3}.
\end{aligned}\right\} \tag{55}$$

An analysis of eqns. (55) shows that the three unbalanced voltage vectors V_a, V_b, and V_c may be resolved into three separate systems of vectors consisting of:

(1) Three equal vectors $\frac{V_a + V_b + V_c}{3}$ which represents a zero-sequence system where $V_{a0} = V_{b0} = V_{c0}$.

(2) A symmetrical three-phase system having positive-phase sequence and represented by V_{a1}, h^2V_{a1}, hV_{a1}, where

$$V_{a1} = \tfrac{1}{3}(V_a + hV_b + h^2V_c)$$

and also

$$h^2V_{a1} = V_{b1},$$

$$hV_{a1} = V_{c1}.$$

(3) A symmetrical three-phase system having negative-phase sequence and represented by V_{a2}, hV_{a2}, h^2V_{a2} where

$$V_{a2} = \tfrac{1}{3}(V_a + h^2V_b + hV_c)$$

and also

$$hV_{a2} = V_{b2},$$

$$h^2V_{a2} = V_{c2}.$$

Equations (55) may thus be written:

$$\left.\begin{aligned} V_a &= V_{a0} + V_{a1} + V_{a2}, \\ V_b &= V_{a0} + h^2 V_{a1} + h V_{a2}, \\ V_c &= V_{a0} + h V_{a1} + h^2 V_{a2}. \end{aligned}\right\} \tag{56}$$

Similarly, a set of equations may be written for unbalanced currents as

$$\left.\begin{aligned} I_a &= I_{a0} + I_{a1} + I_{a2}, \\ I_b &= I_{a0} + h^2 I_{a1} + h I_{a2}, \\ I_c &= I_{a0} + h I_{a1} + h^2 I_{a2}. \end{aligned}\right\} \tag{57}$$

In eqns. (56) and (57) the unbalanced vectors have been expressed in terms of the phase sequence components of the *a* phase. They may equally well be written in terms of the *b* or *c* phase sequence components as follows:

$$\left.\begin{aligned} V_a &= V_{b0} + h V_{b1} + h^2 V_{b2}, \\ V_b &= V_{b0} + V_{b1} + V_{b2}, \\ V_c &= V_{b0} + h^2 V_{b1} + h V_{b2}, \end{aligned}\right\} \tag{58}$$

and also

$$\left.\begin{aligned} V_a &= V_{c0} + h^2 V_{c1} + h V_{c2}, \\ V_b &= V_{c0} + h V_{c1} + h^2 V_{c2}, \\ V_c &= V_{c0} + V_{c1} + V_{c2}. \end{aligned}\right\} \tag{59}$$

15. Solution of symmetrical components

Considering eqns. (56), and making use of the identity $1 + h + h^2 = 0$, adding the equations gives

$$V_a + V_b + V_c = 3V_{a0}.$$

Therefore $$V_{a0} = \tfrac{1}{3}(V_a + V_b + V_c).$$

Also

$$\left.\begin{aligned} V_a &= V_{a0} + V_{a1} + V_{a2}, \\ hV_b &= hV_{a0} + V_{a1} + h^2V_{a2}, \\ h^2V_c &= h^2V_{a0} + V_{a1} + hV_{a2}. \end{aligned}\right\}$$

Adding gives

$$V_a + hV_b + h^2V_c = 3V_{a1}.$$

Therefore $$V_{a1} = \tfrac{1}{3}(V_a + hV_b + h^2V_c).$$

And, finally, to solve for V_{a2},

$$\left.\begin{aligned} V_a &= V_{a0} + V_{a1} + V_{a2}, \\ h^2V_b &= h^2V_{a0} + hV_{a1} + V_{a2}, \\ hV_c &= hV_{a0} + h^2V_{a1} + V_{a2}, \end{aligned}\right\} \qquad (59a)$$

and adding

$$V_a + h^2V_b + hV_c = 3V_{a2}.$$

Therefore $$V_{a2} = \tfrac{1}{3}(V_a + h^2V_b + hV_c).$$

Summarizing the above results we have:

$$\left.\begin{aligned} V_{a0} &= \tfrac{1}{3}(V_a + V_b + V_c), \\ V_{a1} &= \tfrac{1}{3}(V_a + hV_b + h^2V_c), \\ V_{a2} &= \tfrac{1}{3}(V_a + h^2V_b + hV_c). \end{aligned}\right\} \qquad (60)$$

and similarly for currents:

$$\left.\begin{aligned} I_{a0} &= \tfrac{1}{3}(I_a + I_b + I_c), \\ I_{a1} &= \tfrac{1}{3}(I_a + hI_b + h^2I_c), \\ I_{a2} &= \tfrac{1}{3}(I_a + h^2I_b + hI_c). \end{aligned}\right\} \qquad (61)$$

The above equations enable the symmetrical components of the *a*-phase to be calculated and a similar set of equations may be written to determine the *b*- and *c*-phase components, or alternatively they may be calculated using the *a*-phase components as follows:

(i) Zero-phase sequence: $V_{a0} = V_{b0} = V_{c0}$.

(ii) Positive-phase sequence: $V_{b1} = hV_{a1}$,

$$V_{c1} = h^2 V_{a1}.$$

(iii) Negative-phase sequence: $V_{b2} = h^2 V_{a2}$,

$$V_{c2} = hV_{a2}.$$

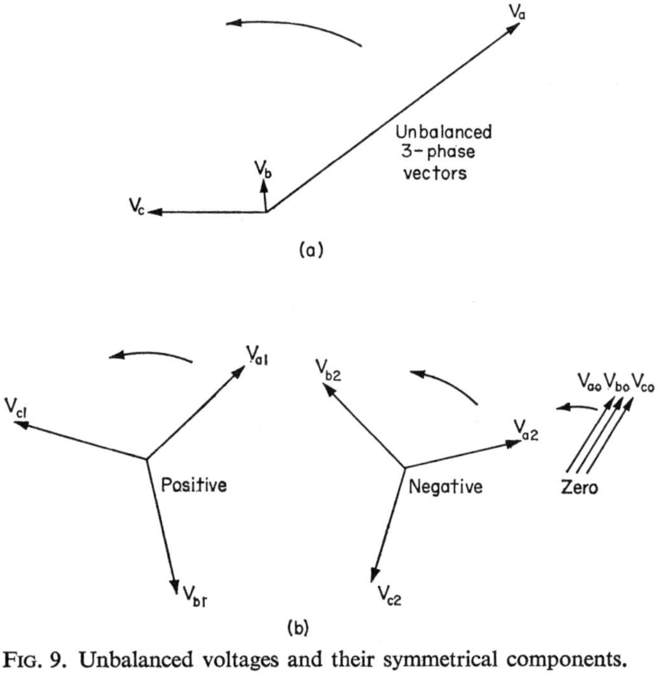

FIG. 9. Unbalanced voltages and their symmetrical components.

In Fig. 9a is shown an unbalanced three-phase system of voltages V_a, V_b, and V_c, and Fig. 9b shows three sets of balanced three-phase vectors which when added together vectorially for each phase result in the unbalanced system. The three balanced systems are the positive-, negative-, and zero-phase sequence components respectively.

16. Sum and difference equations

By writing $h = (-\frac{1}{2} + j\frac{1}{2}\sqrt{3})$ and $h^2 = (-\frac{1}{2} - j\frac{1}{2}\sqrt{3})$ in the expressions for the positive- and negative-sequence components of eqns. (60), the following equations may be obtained:

$$\left.\begin{aligned} V_{a1} &= \tfrac{1}{3}[V_a - \tfrac{1}{2}(V_b + V_c) + j\tfrac{1}{2}\sqrt{3}(V_b - V_c)], \\ V_{a2} &= \tfrac{1}{3}[V_a - \tfrac{1}{2}(V_b + V_c) - j\tfrac{1}{2}\sqrt{3}(V_b - V_c)]. \end{aligned}\right\} \quad (62)$$

Also the expressions for V_b and V_c in eqns. (56) may be rewritten:

$$\left.\begin{aligned} V_b &= V_{a0} - \tfrac{1}{2}(V_{a1} + V_{a2}) - j\tfrac{1}{2}\sqrt{3}(V_{a1} - V_{a2}), \\ V_c &= V_{a0} - \tfrac{1}{2}(V_{a1} + V_{a2}) + j\tfrac{1}{2}\sqrt{3}(V_{a1} - V_{a2}). \end{aligned}\right\} \quad (63)$$

Equations (62) and (63) offer a certain amount of simplification in the numerical calculations since they avoid the multiplication of two complex quantities.

EXAMPLE 7. The line to neutral voltages in a three-phase system are given by

$$V_a = 120\underline{/0^\circ},$$

$$V_b = 150\underline{/330^\circ},$$

$$V_c = 100\underline{/220^\circ}.$$

Determine the symmetrical components.

$$V_a = 120^\circ\underline{/0^\circ} = 120 + j0,$$

$$V_b = 150\underline{/330^\circ} = 129{\cdot}9 - j75,$$

$$V_c = 100\underline{/220^\circ} = -76{\cdot}6 - j64{\cdot}3.$$

Then

$$V_{a0} = \tfrac{1}{3}(V_a + V_b + V_c),$$

$$= \tfrac{1}{3}[(120 + j0) + (129{\cdot}9 - j75) + (-76{\cdot}6 - j64{\cdot}3)]$$

$$= \tfrac{1}{3}[173{\cdot}3 - j139{\cdot}3]$$

$$= 57{\cdot}8 - j46{\cdot}4.$$

There are now three methods by which the positive- and negative-sequence components may be calculated.

Method 1. Use of polar coordinates.

$$V_{a1} = \tfrac{1}{3}[V_a + hV_b + h^2V_c],$$

$$= \tfrac{1}{3}[120\underline{/0^\circ} + 1\underline{/120^\circ} \times 150\underline{/330^\circ} + 1\underline{/240^\circ} \times 100\underline{/220^\circ}],$$

$$= \tfrac{1}{3}[120\underline{/0^\circ} + 150\underline{/90^\circ} + 100\underline{/100^\circ}],$$

$$= \tfrac{1}{3}[(120 + j0) + (0 + j150) + (-17{\cdot}4 + j98{\cdot}5)],$$

$$= \tfrac{1}{3}[102{\cdot}6 + j248{\cdot}5],$$

$$= 34{\cdot}2 + j82{\cdot}8;$$

$$V_{a2} = \tfrac{1}{3}[V_a + h^2V_b + hV_c],$$

$$= \tfrac{1}{3}[120\underline{/0^\circ} + 1\underline{/240^\circ} \times 150\underline{/330^\circ} + 1\underline{/120^\circ} \times 100\underline{/220^\circ}],$$

$$= \tfrac{1}{3}[120\underline{/0^\circ} + 150\underline{/210^\circ} + 100\underline{/340^\circ}],$$

$$= \tfrac{1}{3}[(120 + j0) + (-130 - j75) + (94 - j34{\cdot}2)]$$

$$= \tfrac{1}{3}[84 - j109{\cdot}2]$$

$$= 28 - j36{\cdot}4$$

Method 2. Use of sum and difference equations.

$$(V_b + V_c) = [(129\cdot9 - j75) + (-76\cdot6 - j64\cdot3)],$$
$$= 53\cdot3 - j139\cdot3;$$
$$(V_b - V_c) = [(129\cdot9 - j75) - (-76\cdot6 - j64\cdot3)],$$
$$= 206\cdot5 - j10\cdot7;$$
$$V_{a1} = \tfrac{1}{3}[V_a - \tfrac{1}{2}(V_b + V_c) + j\tfrac{1}{2}\sqrt{3}(V_b - V_c)],$$
$$= \tfrac{1}{3}[(120 + j0) - \tfrac{1}{2}(53\cdot3 - j139\cdot3) + j\tfrac{1}{2}\sqrt{3}(206\cdot5 - j10\cdot7)],$$
$$= \tfrac{1}{3}[102\cdot6 + j248\cdot5],$$
$$= 34\cdot2 + j82\cdot8.$$
$$V_{a2} = \tfrac{1}{3}[V_a - \tfrac{1}{2}(V_b + V_c) - j\tfrac{1}{2}\sqrt{3}(V_b - V_c)],$$
$$= \tfrac{1}{3}[(120 + j0) - \tfrac{1}{2}(53\cdot3 - j139\cdot3) - j\tfrac{1}{2}\sqrt{3}(206\cdot5 - j10\cdot7)]$$
$$= \tfrac{1}{3}[84 - j109\cdot2],$$
$$= 28 - j36\cdot4.$$

Method 3. Graphical method.

In Fig. 10 are shown the geometrical constructions for determining the symmetrical components of the *a*-phase which make use of eqn. (60) from which the constructions are self-evident.

By the above methods the symmetrical components of the *a*-phase have been found where

$$V_{a0} = 57\cdot8 - j46\cdot4 = 74\cdot1 \underline{/321\cdot2^\circ},$$
$$V_{a1} = 34\cdot2 + j82\cdot8 = 89\cdot6 \underline{/67\cdot6^\circ},$$
$$V_{a2} = 28 - j36\cdot4 = 45\cdot9 \underline{/307\cdot6^\circ}.$$

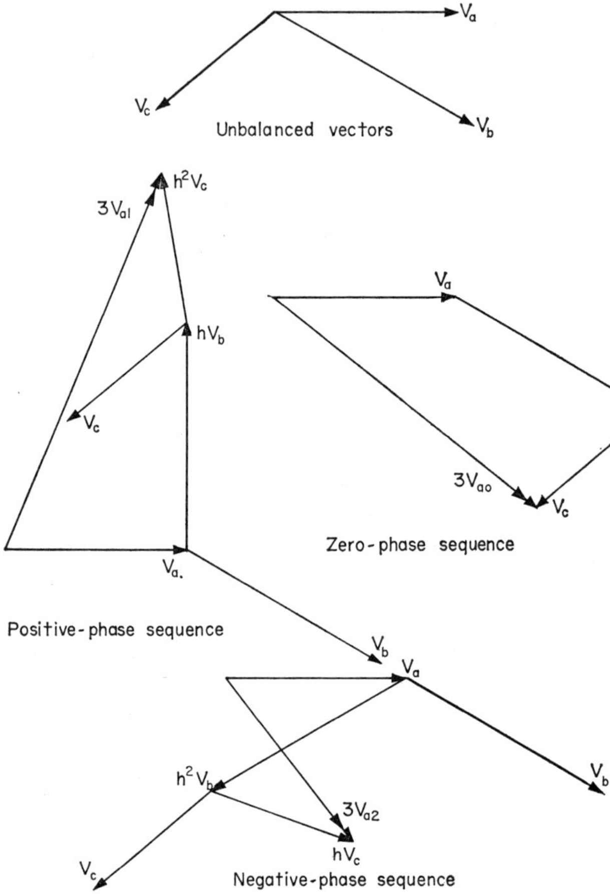

FIG. 10. Example 7.

The components of the b- and c-phases are then given by

$$V_{b0} = V_{a0} = 74\cdot1\ \underline{/321\cdot2^\circ} = 57\cdot8 - j46\cdot4,$$

$$V_{c0} = V_{a0} = 74\cdot1\ \underline{/321\cdot2^\circ} = 57\cdot8 - j46\cdot4,$$

$$V_{b1} = h^2V_{a1} = 89\cdot6\ \underline{/307\cdot6^\circ} = 54\cdot6 - j71\cdot1,$$

$$V_{c1} = hV_{a1} = 89\cdot6\ \underline{/187\cdot6^\circ} = -88\cdot9 - j11\cdot8,$$

$$V_{b2} = hV_{b1} = 45\cdot9\ \underline{/67\cdot6^\circ} = 17\cdot5 + j42\cdot5,$$

$$V_{c2} = h^2V_{b1} = 45\cdot9\ \underline{/187\cdot6^\circ} = -45\cdot5 - j6\cdot1.$$

It is now possible to check the values of the symmetrical components thus obtained by addition to give the original unbalanced vectors V_a, V_b, V_c.

$$\begin{aligned} V_a &= V_{a0} + V_{a1} + V_{a2}, \\ &= (57\cdot8 - j46\cdot4) + (34\cdot2 + j82\cdot8) + (28 - j36\cdot4), \\ &= 120 + j0, \\ &= 120\ \underline{/0^\circ}; \end{aligned}$$

$$\begin{aligned} V_b &= V_{b0} + V_{b1} + V_{b2}, \\ &= (57\cdot8 - j46\cdot4) + (54\cdot6 - j71\cdot1) + (17\cdot5 + j42\cdot5), \\ &= 129\cdot9 - j75, \\ &= 150\ \underline{/330^\circ}; \end{aligned}$$

$$\begin{aligned} V_c &= V_{c0} + V_{c1} + V_{c2}, \\ &= (57\cdot8 - j46\cdot4) + (-88\cdot9 - j11.8) \\ &\qquad + (-45\cdot5 - j6\cdot1), \\ &= -76\cdot6 - j64\cdot3, \\ &= 100\ \underline{/220^\circ}. \end{aligned}$$

17. Zero-sequence currents

In Fig. 11 the three-phase currents I_a, I_b, and I_c are shown to flow in a star-connected load. If the currents are unbalanced they may be resolved into symmetrical components and, furthermore,

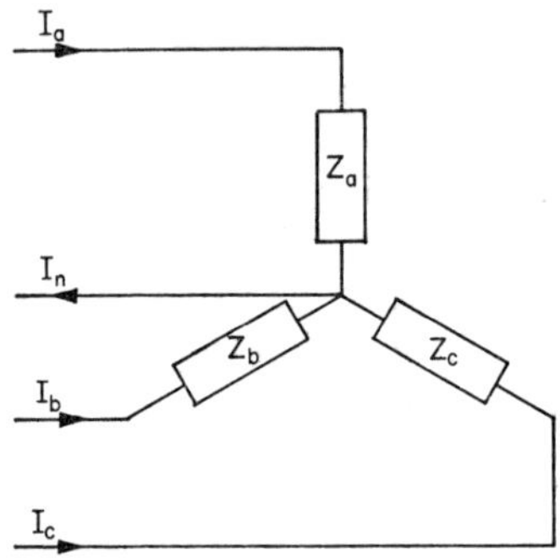

FIG. 11. Unbalanced star-connected lead.

if their vectorial sum is not zero, a current I_n will flow in the neutral conductor where

$$I_n = I_a + I_b + I_c$$

and since, as has been shown previously,

$$I_{a0} = I_{b0} = I_{c0} = \tfrac{1}{3}(I_a + I_b + I_c),$$

then $$I_n = 31_{a0},$$
$$= (I_{a0} + I_{b0} + I_{c0}).$$

Hence the neutral current consists of the sum of the zero-phase sequence components of the currents in the three phases. If, in Fig. 1, there were no neutral connection, then it would be necessary for the vectorial sum of the three-phase currents to be zero and hence there could be no zero-phase sequence component of current. It would, however, be possible for the currents to be unbalanced, in which case only positive- and negative-phase sequence components would exist since their vectorial sum in the three phases would be zero, i.e.

$$(V_{a1} + V_{a2}) + (V_{b1} + V_{b2}) + (V_{c1} + V_{c2}) = 0.$$

If the star-connected load is replaced by three impedances connected in delta, again it would be necessary for the three-phase currents to sum to zero in which case there could be no zero-sequence component in the three lines, although it is possible in certain cases for zero-sequence currents to circulate round the delta.

Generally, it may be stated that zero-sequence currents may only flow in a star-connected load when a neutral connection is present, or in certain circumstances they may circulate round a delta connection but not in the lines.

EXAMPLE 8. A 3-phase, 4-wire system supplies loads which are unequally distributed on the three phases. An analysis of the currents flowing in the direction of the loads in the *R*, *Y* and *B* lines shows that in the *R* line the positive-phase-sequence component of current is $200\underline{/0°}$ A, and the negative-phase-sequence current is $100\underline{/60°}$ A. The total observed current flowing back to the supply in the neutral conductor is $300\underline{/300°}$ A. Calculate the currents in phase and magnitude in the three lines.

Assuming that the 3-phase supply voltages are symmetrical and that the power factor of the load on the *R* phase is $\frac{1}{2}\sqrt{3}$ leading, determine the power factor of the loads on the other two phases.

[I.E.E., June 1960]

Since the sum of the positive- and negative-phase sequence components of current in the three phases must be zero, the residue which flows in the neutral conductor must be the sum of the zero-phase sequence components of the three phases, i.e.

$$I_n = I_{RO} + I_{YO} + I_{BO} = 300\,\underline{/300^\circ},$$

$$I_{RO} = I_{YO} = I_{BO} = \tfrac{1}{3}\,I_n = 100\,\underline{/300^\circ}\text{ A.}$$

Now
$$\begin{aligned} I_R &= I_{RO} + I_{R1} + I_{R2}, \\ &= 100\,\underline{/300^\circ} + 200\,\underline{/0^\circ} + 100\,\underline{/60^\circ}, \\ &= (50 - j86{\cdot}6) + (200 + j0) + (50 + j86{\cdot}6), \\ &= 300 + j0, \\ &= 300\,\underline{/0^\circ}\text{ A;} \end{aligned}$$

also
$$\begin{aligned} I_Y &= I_{RO} + h^2 I_{R1} + h I_{R2}, \\ &= 100\,\underline{/300^\circ} + 200\,\underline{/240^\circ} + 100\,\underline{/180^\circ}, \\ &= (50 - j86{\cdot}6) + (-\,100 - j173{\cdot}2) \\ &\qquad + (-\,100 + j0), \\ &= -\,150 - j259{\cdot}8, \\ &= 300\,\underline{/240^\circ}\text{ A.} \end{aligned}$$

And
$$\begin{aligned} I_B &= I_{RO} + h I_{R1} + h^2 I_{R2}, \\ &= 100\,\underline{/300^\circ} + 200\,\underline{/120^\circ} + 100\,\underline{/300^\circ}, \\ &= (50 - j86{\cdot}6) + (-\,100 + j173{\cdot}2) \\ &\qquad + (50 - j86{\cdot}6), \\ &= 0. \end{aligned}$$

Figure 12 shows the vector relationship between the phase values of voltage and current. Since the power factor of the load in the R-phase is $\frac{1}{2}\sqrt{3}$, leading the angle between vectors V_R and I_R will be

$$\cos^{-1}\tfrac{1}{2}\sqrt{3} = 30^\circ$$

with I_R leading V_R. As the voltages are balanced their vectors will be 120° apart, as shown, giving the power factor angle of the load in the Y-phase as 30° lead.

Hence p.f. of load on Y-phase $= \cos 30° = \frac{1}{2}\sqrt{3}$ leading.

As the B-phase current is zero the power factor is indeterminate.

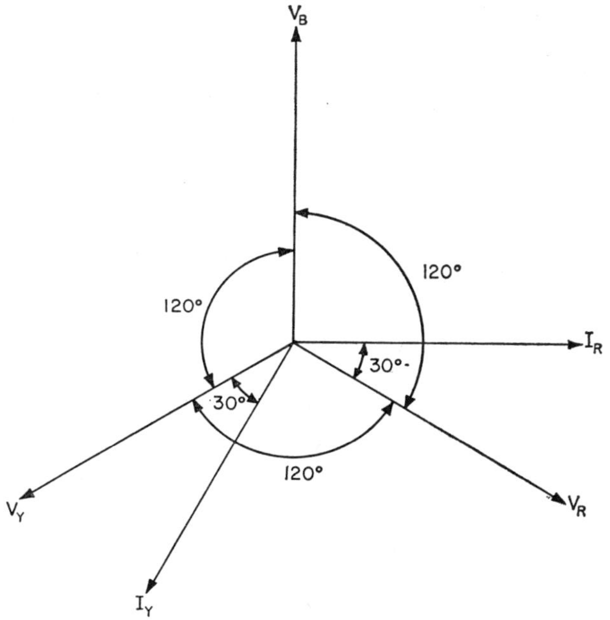

FIG. 12. Example 8.

18. Line-to-line voltages

The line-to-line voltages of a three-phase system may be expressed in terms of the phase (i.e. line-to-neutral) voltages, and hence the phase sequence components of the phase voltages, where

$$
\left.\begin{aligned}
V_{ab} = V_a - V_b &= (V_{a0} + V_{a1} + V_{a2}) - (V_{a0} + h^2 V_{a1} + hV_{a2}),\\
&= (1 - h^2)\, V_{a1} + (1 + h)\, V_{a2};\\
V_{bc} = V_b - V_c &= (V_{a0} + h^2 V_{a1} + hV_{a2})\\
&\qquad - (V_{a0} + hV_{a1} + h^2 V_{a2}),\\
&= h^2 (1 - h^2)\, V_{a1} + h\,(1 - h)\, V_{a2};\\
V_{ca} = V_c - V_a &= (V_{a0} + hV_{a1} + h^2 V_{a2})\\
&\qquad - (V_{a0} + V_{a1} + V_{a2}),\\
&= h\,(1 - h^2)\, V_{a1} + h^2 (1 - h)\, V_{a2}.
\end{aligned}\right\} \tag{64}
$$

It is thus apparent that the line voltages are wholly independent of the zero-sequence components of the phase voltages, and from eqn. (64) it may be seen that the vectorial sum of the line voltages must sum to zero, and hence if shown graphically the three vectors representing the line voltages would form a closed triangle.

Since the line voltages are independent of the zero-sequence component it is obviously not possible to determine this component from a knowledge only of the line voltages. It is, however, possible to derive the positive- and negative-sequence components of the phase voltages, for multiplying eqns. (64) by 1, h, and h^2 respectively gives

$$
\begin{aligned}
V_{ab} + hV_{bc} + h^2 V_{ca} &= 3\,(1 - h^2)\, V_{a1},\\
V_{a1} &= \frac{1}{3\,(1 - h^2)}\,(V_{ab} + hV_{bc} + h^2 V_{ca}),\\
&= \tfrac{1}{9}\,(1 - h)\,(V_{ab} + hV_{bc} + h^2 V_{ca}).
\end{aligned} \tag{65}
$$

Also multiplying eqns. (77) by 1, h^2, and h respectively gives

$$V_{ab} + h^2 V_{bc} + hV_{ca} = 3\,(1 - h)\, V_{a2}.$$

Therefore

$$
\begin{aligned}
V_{a2} &= \frac{1}{3\,(1 - h)}\,(V_{ab} + h^2 V_{bc} + hV_{ca}),\\
&= \tfrac{1}{9}\,(1 - h^2)\,(V_{ab} + h^2 V_{bc} + hV_{ca}).
\end{aligned} \tag{66}
$$

This means now that, since only the positive- and negative-sequence components of the phase voltages are determinate from the line voltages, the phase voltages may only be derived from the line voltages when the phase voltages are without residue.

If they contain zero-sequence components then their values are indeterminate.

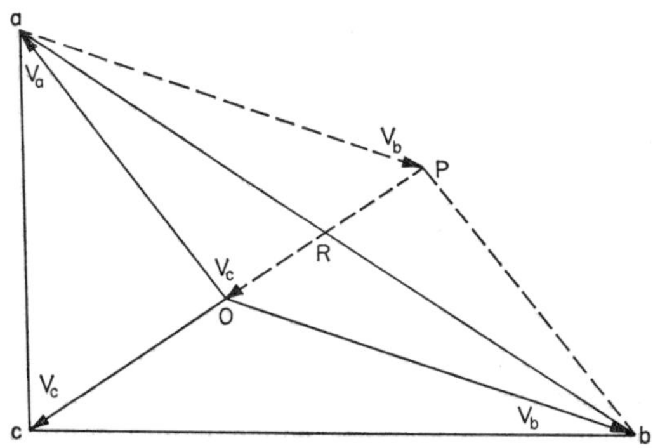

FIG. 13. Graphical determination of phase voltages.

19. Graphical determination of phase voltages from line voltages when zero-sequence component is absent

The three-phase voltages V_a, V_b, and V_c are represented in Fig. 13 by *Oa*, *Ob*, and *Oc* respectively. The sides of the triangle *abc* will then be such that the line voltages V_{ab}, V_{bc}, and V_{ca} will be represented by vectors *ba*, *cb*, and *ac*.

If the phase voltages contain no zero-phase sequence components they will be without residue and

$$V_a + V_b + V_c = 0.$$

Constructing the vectors *aP*, equal to V_b, and *PO*, equal to V_c, as shown, *PO* will intersect *ab* at *R*.

From the geometry of the figure $Pb = Oa$ making $OaPb$ a parallelogram from which $aR = Rb$ and also cOP must be a straight line since PO and Oc both represent the same vector V_c. Hence cO extended must bisect ab. Similar constructions on Oc and Ob can be made to show that both bO and aO extended will bisect ac and bc respectively. Thus the point O must be the centroid of the triangle abc.

If now only the line voltages are known, and it is also known that the phase voltages are without residue, then by constructing the triangle abc to represent the line voltages, the phase voltages may be found from determining the centroid of the triangle as in the diagram.

There is only one possible solution for the phase voltages when the line voltages are known and the zero-sequence component is absent.

If a zero-sequence component is present in the phase voltages there is no method by which it may be determined from the line values. The above method, however, could be used to determine three equivalent phase voltages which would contain the same positive- and negative-sequence components, and from the equivalent phase voltages these two components could then be obtained.

20. Alternative graphical method to determine zero-phase sequence voltage

When the three-phase values of voltage do not sum to zero vectorially, then a zero-phase sequence component of voltage must be present. Three such phase voltages, V_a, V_b, and V_c, are represented by Oa, Ob, and Oc in Fig. 14 so that the line voltages V_{ab}, V_{bc}, and V_{ca} will be represented by ba, cb, and ac respectively. If the centroid O' of triangle abc is now determined, the same line voltages could result from the three-phase voltages V'_a, V'_b, and V'_c which are without residue as shown in § 19. Now since V'_a, V'_b, and V'_c contain only positive- and negative-phase sequence components and are the only possible solution for such phase voltages having the line voltages shown, they must contain the same positive- and negative-phase sequence components as the original phase voltages V_a, V_b, and V_c. The following relationships are then evident:

$$V_a' = V_{a1} + V_{a2},$$

$$V_b' = V_{b1} + V_{b2},$$

$$V_c' = V_{c1} + V_{c2}.$$

Now $\quad 00' = V_a - V_a',$

$$= (V_{a0} + V_{a1} + V_{a2}) - (V_{a1} + V_{a2}),$$

$$= V_{a0},$$

and so a vector drawn to join O and O', the centroid of the triangle abc, would represent the zero-phase sequence component of the phase voltages.

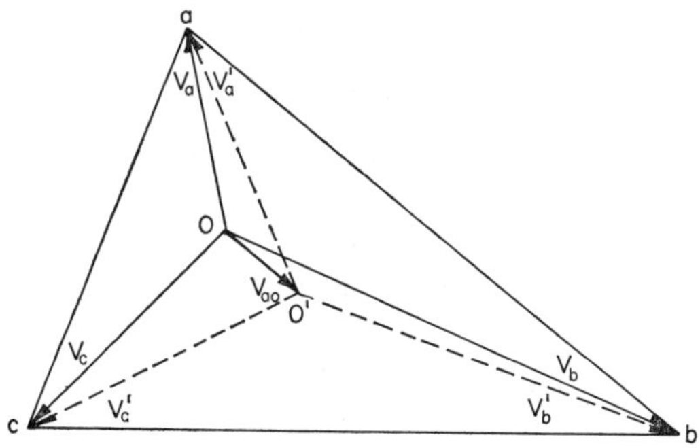

FIG. 14. Graphical determination of zero-phase sequence voltage.

21. Alternative graphical method to determine positive- and negative-phase sequence voltages from phase voltages

This construction (Fig. 15) may be used only when either the phase voltages do not contain a zero-phase sequence component or if the zero-phase sequence component has been subtracted from the phase values.

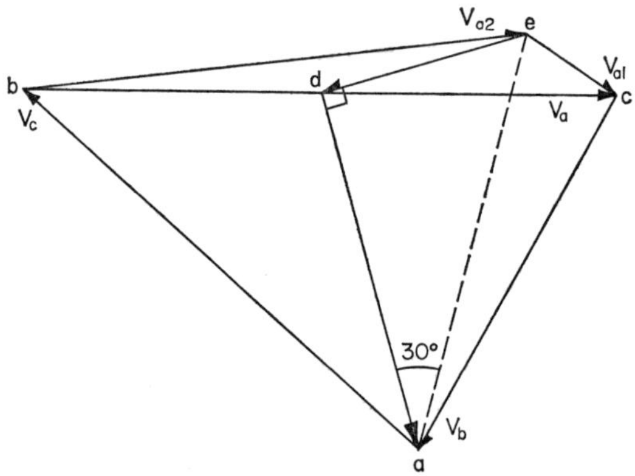

FIG. 15. Graphical determination of positive- and negative-phase sequence voltages from phase voltages.

For the positive phase sequence component

$$V_{a1} = \tfrac{1}{3}(V_a + hV_b + h^2V_c)$$

and in the absence of the zero-phase sequence component

$$V_a + V_b + V_c = 0,$$

$$\begin{aligned} V_{a1} &= \tfrac{1}{3}[V_a + hV_b + h^2(-V_a - V_b)], \\ &= \tfrac{1}{3}[(1 - h^2)V_a + (h - h^2)V_b], \\ &= \tfrac{1}{3}[(\tfrac{3}{2} + j\tfrac{1}{2}\sqrt{3})V_a + j\sqrt{(3)}\,V_b], \\ &= \frac{V_a}{2} + \frac{1}{\sqrt{3}}\left(\frac{V_a}{2} + V_b\right)\underline{/90^\circ}. \end{aligned} \tag{67}$$

Similarly, for the negative phase sequence component,

$$\begin{aligned} V_{a2} &= \tfrac{1}{3}(V_a + h^2V_b + hV_c), \\ &= \tfrac{1}{3}[V_a + h^2V_b + h(-V_a - V_b)], \\ &= \tfrac{1}{3}[(1-h)V_a + (h^2 - h)V_b], \\ &= \tfrac{1}{3}[(\tfrac{3}{2} - j\tfrac{1}{2}\sqrt{3})V_a - j\sqrt{(3)}V_b], \\ &= \frac{V_a}{2} - \frac{1}{\sqrt{3}}\left(\frac{V_a}{2} + V_b\right)\underline{/90^\circ}. \end{aligned} \tag{68}$$

The above relationships given in eqns. (67) and (68) may be determined graphically by the following method.

The three unbalanced voltage vectors V_a, V_b, and V_c are drawn to form a closed triangle *abc*.

ad is then drawn to bisect *bc* so that

$$bd = dc = \frac{V_a}{2}.$$

Then, vectorially,

$$da = \frac{V_a}{2} + V_b.$$

A triangle *ade* is now constructed on *ad* such that

$$\angle\, ead = 30^\circ,$$

and

$$\angle\, ade = 90^\circ$$

Now, in such a triangle, the ratio of the length of the sides is given by

$$ad : de : ea = 1 : \sqrt{3} : 2.$$

Hence, in magnitude,

$$ed = \frac{da}{\sqrt{3}},$$

and, since the angle between *da* and *ed* is 90°, the vector relationship is

$$ed = \frac{da}{\sqrt{3}} \underline{/90^\circ},$$

$$= \frac{1}{\sqrt{3}} \left[\frac{V_a}{2} + V_b\right] \underline{/90^\circ}.$$

The vector *ec* is now given by

$$ec = ed + dc,$$

$$= \frac{1}{\sqrt{3}} \left[\frac{V_a}{2} + V_b\right] \underline{/90^\circ} + V_a,$$

$$= V_{a1},$$

and the vector *be* is given by

$$be = bd - ed,$$

$$= \frac{V_a}{2} - \frac{1}{\sqrt{3}} \left[\frac{V_a}{2} + V_b\right] \underline{/90^\circ}.$$

22. Graphical determination of positive- and negative-phase sequence voltages from line voltages

The positive-phase sequence component V_{a1} of the *a*-phase is given in terms of the three line voltages by eqn. (65). Since the vectorial sum of the three line voltages must be zero, V_{a1} may be expressed in terms of the two-line voltages V_{ab} and V_{bc} alone.

$$\begin{aligned} V_{a1} &= \tfrac{1}{9}(1 - h)\,[V_{ab} + hV_{bc} + h^2V_{ca}], \\ &= \tfrac{1}{9}(1 - h)\,[V_{ab} + hV_c + h^2(-V_{ab} - V_{bc})], \\ &= \tfrac{1}{9}(1 - h)\,[(1 - h^2)\,V_{ab} + (h - h^2)\,V_{bc}], \\ &= \tfrac{1}{9}\,[3V_{ab} - 3h^2V_{bc}], \\ &= \tfrac{1}{3}\,[V_{ab} - h^2V_{bc}], \\ &= \tfrac{1}{3}\,[V_{ab} + V_{bc}\,\underline{/60^\circ}]. \end{aligned} \tag{69}$$

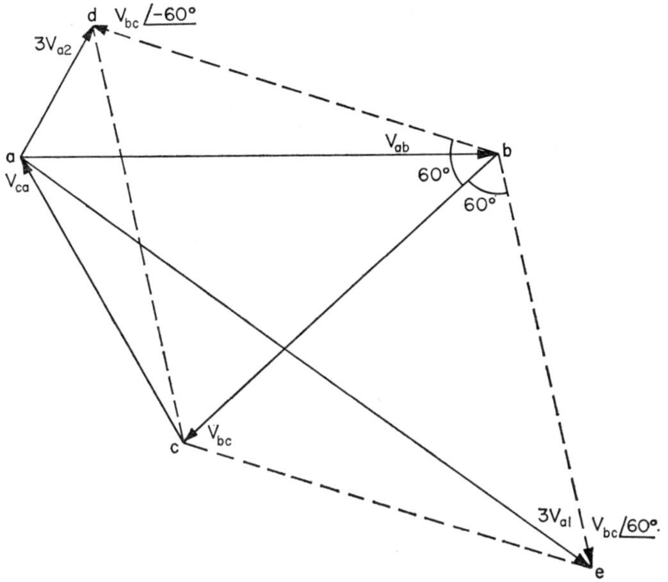

FIG. 16. Graphical determination of positive- and negative-phase sequence voltages from line voltages.

If, now, the three line voltages are drawn to represent the sides of a triangle *abc* as in Fig. 16 and an equilaterial triangle *bce* is constructed on *bc*, then a vector *be* will represent $V_{bc}\underline{/60°}$. Joining *a* to *e*, *ae* will then be represented vectorially by

$$\begin{aligned} ae &= ab + be, \\ &= V_{ab} + V_{bc}\,\underline{/60°}, \\ &= 3V_{a1}. \end{aligned}$$

Also, from eqn. (66), the negative-phase sequence component V_{a2} may be derived in terms of the line voltages V_{ab} and V_{bc} alone.

$$V_{a2} = \tfrac{1}{9}(1 - h^2)[V_{ab} + h^2V_{bc} + hV_{ca}],$$

$$= \tfrac{1}{9}(1 - h^2)[V_{ab} + h^2V_{bc} + h(-V_{ab} - V_{bc})],$$

$$= \tfrac{1}{9}(1 - h^2)[(1 - h)V_{ab} + (h^2 - h)V_{bc}],$$

$$= \tfrac{1}{9}[3V_{ab} - 3hV_{bc}],$$

$$= \tfrac{1}{3}[V_{ab} - hV_{bc}],$$

$$= \tfrac{1}{3}[V_{ab} + V_{bc}\underline{/-60^\circ}]. \qquad (70)$$

Constructing an equilateral triangle *bed* on *be* will give a vector *bd* to represent $V_{bc}\underline{/-60^\circ}$. Joining *a* to *d* now gives a vector *ad* where vectorially

$$ad = ab + bd,$$

$$= V_{ab} + V_{bc}\underline{/-60^\circ},$$

$$= 3V_{a2}.$$

23. Unbalance factor

The ratio of the negative- to the positive-phase sequence component of current in an unbalanced three-phase system is known as the unbalance factor, and the ratio of the negative- to the positive-phase sequence component of voltage as the unsymmetry factor. Generally, the amount of unbalance in the voltage and current will be different and so each must be calculated separately.

The range of values for the unbalance and unsymmetry factors is between zero and infinity from when the negative-phase sequence component is absent to when the positive-phase sequence component is zero. Thus in a balanced three-phase sustem the unbalance and unsymmetry factors would both be zero.

Exercises

1. Solve Example 7 graphically using the methods given in §§ 20 and 21.

2. In a 3-phase 4-wire system the currents in the R, Y and B lines under abnormal conditions of loading were as follows:

$$I_r = 100 \underline{/30^\circ}\,\text{A}, \quad I_y = 50 \underline{/300^\circ}\text{A}, \quad I_b = 30 \underline{/180^\circ}\,\text{A}$$

Calculate the positive, negative and zero-phase-sequence currents in the R line, and the return current in the neutral conductor. [I.E.E., November 1960]

3. The magnitude of line voltages V_{ab}, V_{bc}, and V_{ca}, of a three-phase star-connected system are found to be 1000, 1500, and 750 V respectively. Determine the unsymmetry factor.

4. Under fault conditions in a 3-phase system, the following currents were recorded in the R, Y and B lines:

$$I_R = 3000\underline{/0^\circ}\,\text{A}, \quad I_Y = 2000\underline{/270^\circ}\,\text{A}, \quad I_B = 1000\underline{/120^\circ}\,\text{A}$$

Calculate the values of the positive, negative and zero phase sequence components. [I.E.E., 1957]

CHAPTER 3

The Properties of Symmetrical Components

IT HAS been demonstrated in the preceding chapter how, given unbalanced three-phase voltages or currents, they may be resolved into three separate component sets of three-phase vector systems—positive-, negative-, and zero-phase sequence components. Before proceeding further, it is now necessary to investigate some of the properties of symmetrical components, and to find methods which will simplify their manipulation, and so enable them to be applied later in the solution of problems involving unbalanced faults.

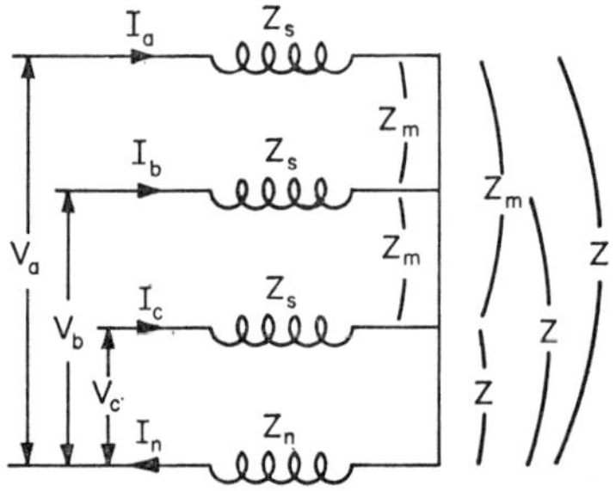

FIG. 17. Symmetrical static network.

24. Phase sequence currents and voltages in a symmetrical network

The general case of a symmetrical static network is represented in Fig. 17, in which the three star-connected impedances Z_s are equal, the mutual impedances between lines are each equal to Z_m.

the mutual impedances between line and neutral are each equal to Z, and the neutral impedance is Z_n. In order to see how such a network behaves, it is necessary to derive the relationships between the phase sequence currents and voltages. Let the three unbalanced line to neutral voltages V_a, V_b, and V_c, be applied to the network so causing currents I_a, I_b, and I_c to flow in the three lines and a current I_n in the neutral.

Equating the voltage drops round the network gives

$$\begin{aligned} V_a &= I_a Z_s + (I_b + I_c) Z_m - I_n Z + I_n Z_n - (I_a + I_b + I_c) Z, \\ &= I_a(Z_s - Z) + I_b(Z_m - Z) + I_c(Z_m - Z) + I_n (Z_n - Z), \\ &= I_a(Z_s - Z) + I_b(Z_m - Z) + I_c(Z_m - Z) \\ &\quad + (I_a + I_b + I_c)(Z_n - Z), \\ &= I_a(Z_s + Z_n - 2Z) + I_b(Z_m + Z_n - 2Z) \\ &\quad + I_c(Z_m + Z_n - 2Z), \end{aligned}$$

and, similarly,

$$\begin{aligned} V_b &= I_a(Z_m + Z_n - 2Z) + I_b(Z_s + Z_n - 2Z) \\ &\quad + I_c(Z_m + Z_n - 2Z), \end{aligned}$$

$$\begin{aligned} V_c &= I_a(Z_m + Z_n - 2Z) + I_b(Z_m + Z_n - 2Z) \\ &\quad + I_c(Z_s + Z_n - 2Z). \end{aligned}$$

The symmetrical components of the applied voltages may now be determined, where

$$\begin{aligned} V_{a0} &= \tfrac{1}{3}(V_a + V_b + V_c), \\ &= \tfrac{1}{3}(I_a + I_b + I_c)(Z_s + 3Z_n + 2Z_m - 6Z), \\ &= I_{a0}(Z_s + 3Z_n + 2Z_m - 6Z); \end{aligned}$$

$$
\begin{aligned}
V_{a1} &= \tfrac{1}{3}(V_a + hV_b + h^2V_c),\\
&= \tfrac{1}{3}\{[I_a(Z_s + Z_n - 2Z) + (I_b + I_c)(Z_m + Z_n - 2Z)]\\
&\quad + h[(I_a + I_c)(Z_m + Z_n - 2Z) + I_b(Z_s + Z_n - 2Z)]\\
&\quad + h^2[(I_a + I_b)(Z_m + Z_n - 2Z) + I_c(Z_s + Z_n - 2Z)]\},\\
&= \tfrac{1}{3}\{(I_a + hI_b + h^2I_c)[(Z_s + Z_n - 2Z)\\
&\quad + h(Z_m + Z_n - 2Z) + h^2(Z_m + Z_n - 2Z)]\},\\
&= I_{a1}[Z_s + Z_n(1 + h + h^2) - 2Z(1 + h + h^2)\\
&\quad + Z_m(h + h^2)],\\
&= I_{a1}(Z_s - Z_m);\\
V_{a2} &= \tfrac{1}{3}(V_a + h^2V_b + hV_c),\\
&= \tfrac{1}{3}\{[I_a(Z_s + Z_n - 2Z) + (I_b + I_c)(Z_m + Z_n - 2Z)]\\
&\quad + h^2[(I_a + I_c)(Z_m + Z_n - 2Z) + I_b(Z_s + Z_n - 2Z)]\\
&\quad + h[(I_a + I_b)(Z_m + Z_n - 2Z) + I_c(Z_s + Z_n - 2Z)]\},\\
&= \tfrac{1}{3}\{(I_a + h^2I_b + hI_c)[(Z_s + Z_n - 2Z)\\
&\quad + h(Z_m + Z_n - 2Z) + h^2(Z_m + Z_n - 2Z)]\},\\
&= I_{a2}[Z_s + Z_n(1 + h + h^2) - 2Z(1 + h + h^2)\\
&\quad + Z_m(h + h^2)],\\
&= I_{a2}(Z_s - Z_m).
\end{aligned}
$$

Summarizing the above results gives

$$
\left.
\begin{aligned}
V_{a0} &= I_{a0}(Z_s + 3Z_n + 2Z_m - 6Z),\\
&= I_{a0}Z_0;\\
V_{a1} &= I_{a1}(Z_s - Z_m),\\
&= I_{a1}Z_1;\\
V_{a2} &= I_{a2}(Z_s - Z_m),\\
&= I_{a2}Z_2,
\end{aligned}
\right\}
\tag{71}
$$

where $Z_0 = (Z_s + 3Z_n + 2Z_m - 6Z)$
$= \text{zero-sequence impedance;}$

$Z_1 = (Z_s - Z_m) = \text{positive-sequence impedance;}$

$Z_2 = (Z_s - Z_m) = \text{negative-sequence impedance.}$

Inspection of eqn. (71) shows that there is no interaction between voltages and currents of different phase sequences. Zero-phase sequence voltages produce only zero-phase sequence currents, positive-phase sequence voltages produce only positive-phase sequence currents, and negative-phase sequence voltages produce only negative-phase sequence currents. This, in general, is true for all symmetrical static networks, and, indeed, is also true for symmetrical networks containing rotating machines provided that they are symmetrically wound. Although Fortescue has given a rigorous proof showing that this is so, the above example serves to illustrate the point, and the result is very important since it leads to considerable simplification when solving networks for the condition of unsymmetrical faults.

The impedances Z_0, Z_1, and Z_2 are defined as the zero-, positive-, and negative-sequence impedances, and are the impedances encountered by the corresponding zero-, positive-, and negative-phase sequence currents. In general, for static circuits, the positive- and negative-sequence impedances will be equal, but different from the zero-sequence impedance as shown in this case, whereas in the case of rotating machines, all three impedances will normally have different values.

EXAMPLE 9. An unbalanced star-connected 3-phase supply, having negligible internal impedance, is connected by three conductors to three identical voltmeters also in star connection. Each voltmeter has a resistance of 10,000 ohms and negligible reactance. The supply voltages of the three phases from line to neutral are, respectively,

$$E_R = 100\,\underline{/0^\circ}, \quad E_Y = 200\,\underline{/270^\circ}, \text{ and } \quad E_B = 100\,\underline{/120^\circ} \text{ volts.}$$

Using the method of symmetrical components determine either graphically or algebraically the reading of the voltmeter connected to the yellow line. [I.E.E., June 1958]

(a) *Algebraic method.* The circuit may be represented as shown in Fig. 18 with the line to neutral voltages star connected, where

$$E_R = 100 \underline{/0^\circ} = 100 + j0,$$

$$E_Y = 200 \underline{/270^\circ} = 0 - j200,$$

$$E_B = 100 \underline{/120^\circ} = -50 + j86{\cdot}6.$$

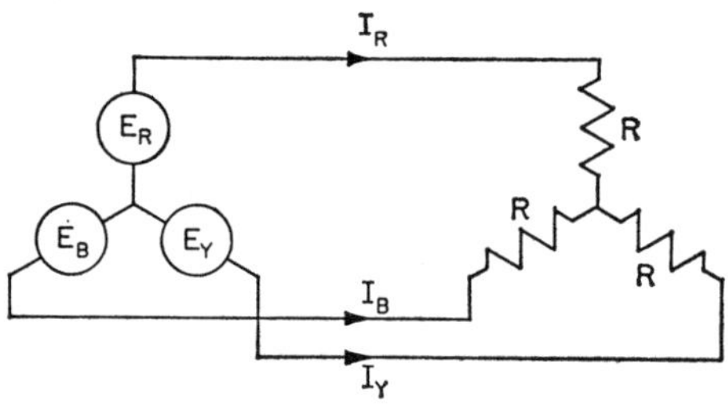

FIG. 18. Example 9.

The three voltmeters are shown as three identical star-connected resistors R.

Let the three line currents be I_R, I_Y, and I_B, then, since there is no neutral connection between the two star points,

$$I_R + I_Y + I_B = 0$$

and it follows that

$$I_{R0} = I_{Y0} = I_{B0} = 0.$$

The voltage across the voltmeter in the yellow line

$$= I_Y R = R\,(I_{Y0} + I_{Y1} + I_{Y2}),$$

$$= R\,(I_{Y1} + I_{Y2}) \quad \text{since } I_{Y0} = 0.$$

Now, since the three voltmeters form a symmetrical network, the phase sequence voltage drops are due to the corresponding phase sequence currents.

Therefore
$$I_Y R = I_{Y1} R + I_{Y2} R,$$
$$= E_{Y1} + E_{Y2},$$

and since
$$\left.\begin{aligned} E_{Y1} &= \tfrac{1}{3}(E_Y + hE_B + h^2 E_R), \\ E_{Y2} &= \tfrac{1}{3}(E_Y + h^2 E_B + hE_R), \end{aligned}\right\} \qquad (71a)$$

then
$$\begin{aligned} I_Y R &= E_{Y1} + E_{Y2}, \\ &= \tfrac{1}{3}[2E_Y + (h + h^2)(E_R + E_B)], \\ &= \tfrac{1}{3}[2E_Y - (E_R + E_B)], \\ &= \tfrac{1}{3}[-j400 + 50 - j86{\cdot}6 - 100], \\ &= -16{\cdot}67 - j162{\cdot}2, \\ &= 163 \underline{/264{\cdot}1^\circ}. \end{aligned}$$

The voltmeter in the yellow line will read the magnitude of this voltage, i.e. 163 V.

(b) *Graphical method.* If the three-phase voltage vectors E_R, E_Y, and E_B, are drawn to scale to give the triangle RYB as in Fig. 19, the sides YR, BY, and RB will then represent the line voltage vectors V_{RY}, V_{YB}, and V_{BR}, respectively. These line voltages will appear across the voltmeter network. Since there is no neutral connection to the star point of the voltmeters there can be no zero-phase sequence current, and hence no zero-phase sequence voltage, across each of the voltmeters. Thus only positive- and negative-phase sequence voltages will appear, and so from § 20 the voltages appearing across the voltmeters are found by determining the centroid O' of the triangle RYB.

The magnitude of $O'Y$ then represents the voltage across the voltmeter in the yellow line, and by measurement may be shown to be equal to 163 V.

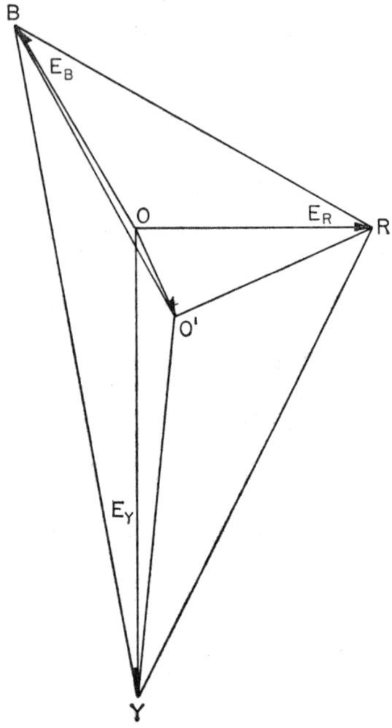

FIG. 19. Example 9—Graphical solution.

Vectors $O'B$ and $O'R$ would also represent the readings of the voltmeters in the blue and red lines respectively.

It is also interesting to note that the vector OO' would also represent the negative-phase sequence voltage, which is the voltage between the two star points of Fig. 18.

(c) *Further consideration.* Further consideration of the above problem shows that, since there is no interaction between the currents and voltages of different phase sequence, the problem may be regarded as one in which the separate phase sequence components of voltage are superimposed on one another and applied to the balanced network of voltmeters as in Fig. 20. So far as the positive- and negative-phase sequence components are concerned, the

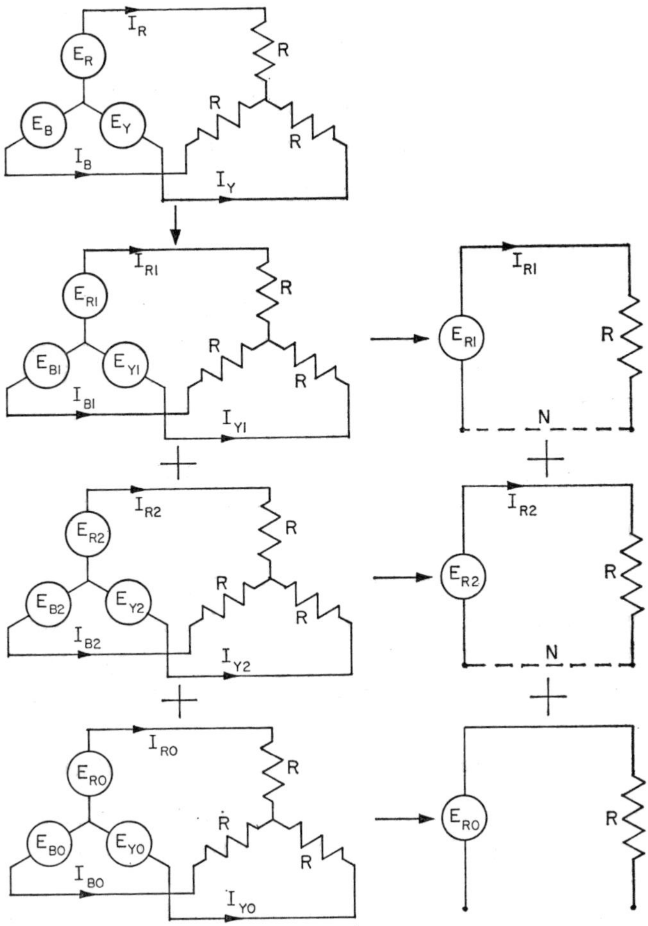

FIG. 20. Phase sequence networks.

corresponding equivalent circuits are similar to the normal balanced three-phase case except for the opposite phase rotation of the negative-phase sequence voltages. In both cases a solution is possible using the normal method adopted for a balanced three-phase network and considering one single phase alone with an assumed neutral connection between the two star points. The positive- and negative-phase sequence currents are then given by $\frac{E_{a1}}{R}$ and $\frac{E_{a2}}{R}$ for the a phase where the equivalent positive and negative sequence impedances are

$$Z_1 = Z_2 = R.$$

For the zero-sequence network the three-phase voltages are equal and in phase with one another, and since there is no connection between the two star points no zero-phase sequence current may flow. Thus when simplifying this three-phase network to one of single phase an open circuit must be shown between the two star points to represent an infinite zero-sequence impedance so that

$$Z_0 = \infty \quad \text{and} \quad I_{a0} = 0.$$

It is thus evident that the complete problem may be regarded as the superimposition of three separate networks each having its own particular set of phase sequence quantities.

25. Generated e.m.f. of a three-phase generator

The voltage which appears across the terminals of a single phase of a three-phase a.c. generator on open circuit is referred to as the phase e.m.f., and is denoted by the symbol E. Although the windings may be connected in either star or delta, the star connection is the more usual and due to the symmetrical arrangements of the windings, and the fact that there are an equal number of turns per phase, the three vectors representing the e.m.f.s will be spaced 120° apart and of equal magnitude. This being so, the three-phase e.m.f.s may be represented by

$$E_a,$$

$$E_b = h^2E_a,$$

$$E_c = hE_a.$$

The phase sequence components of e.m.f. are then:

$$E_{a0} = \tfrac{1}{3}(E_a + E_b + E_c) = \tfrac{1}{3}\, E_a(1 + h^2 + h) = 0,$$

$$E_{a1} = \tfrac{1}{3}(E_a + hE_b + h^2E_c) = \tfrac{1}{3}\, E_a(1 + 1 + 1) = E_a,$$

$$E_{a2} = \tfrac{1}{3}(E_a + h^2E_b + hE_c) = \tfrac{1}{3}\, E_a(1 + h + h^2) = 0,$$

and, as would be expected, it is seen that the only e.m.f.s generated are of positive-phase sequence. Neither negative- nor zero-phase sequence e.m.f.s can be generated in a machine which generates balanced three-phase voltages.

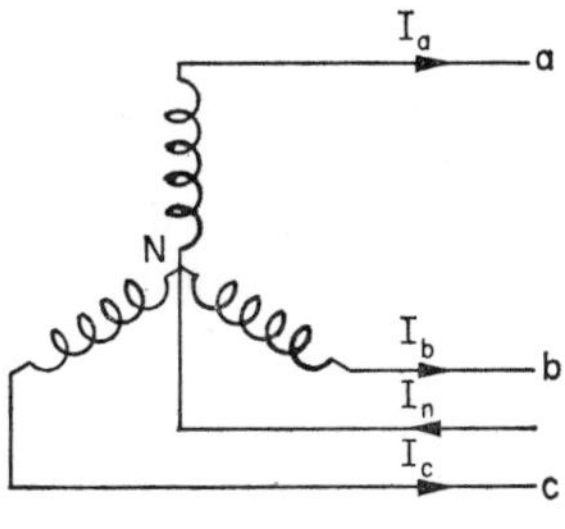

FIG. 21. Neutral current.

26. Neutral current

In the absence of a neutral connection in a three-phase circuit the vectorial sum of the line currents must be zero in accordance with Kirchhoff's first law. When a neutral connection is present, any residual current will flow in the neutral circuit as shown in Fig. 21 where

$$I_n = I_a + I_b + I_c.$$

Expressing I_a, I_b, and I_c in terms of their symmetrical components gives

$$I_n = (I_{a0} + I_{a1} + I_{a2}) + (I_{a0} + h^2 I_{a1} + hI_{a2}) + (I_{a0} + hI_{a1} + h^2 I_{a2}),$$

$$= 3I_{a0} + I_{a1}(1 + h + h^2) + I_{a2}(1 + h + h^2),$$

$$= 3I_{a0}. \tag{72}$$

The neutral current is thus equal to three times the zero-phase sequence component of current and is the sum of the zero-phase sequence currents flowing in each of the three lines, i.e.

$$I_n = I_{a0} + I_{b0} + I_{c0}. \tag{73}$$

Zero-phase sequence currents can only flow when a neutral connection is present since their vectorial sum in the three lines cannot equate to zero as each of their line values is in phase. On the other hand positive- and negative-phase sequence components of current cannot appear in the neutral connection as their resultants in the three lines are zero in each case.

27. Generator with neutral earthed through an impedance

A star-connected generator having its star point solidly earthed without an impedance in the neutral connection will maintain earth potential at the star point. With an impedance connected in the neutral, and unbalanced loading on the generator, current will flow through the neutral impedance. This will cause the star point to assume a potential other than that of earth. Figure 22 shows such an arrangement.

Now if V'_a, V'_b, and V'_c are assumed to be the phase potentials to earth at the neutral point, then

$$V'_a = V'_b = V'_c = V_n.$$

Considering the a-phase alone,

$$V'_a = V_n = -I_n Z_n = -3I_{a0}(Z_n). = -I_{ao}\,(3Z_n).$$

Also $\quad V'_{a0} = -I_{a0}\, Z'_0.$

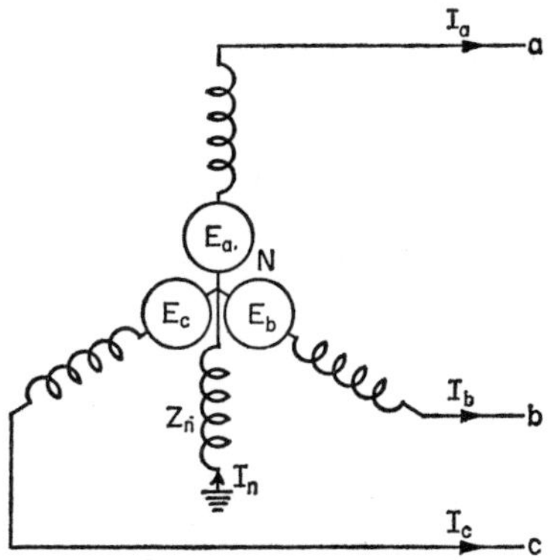

FIG. 22. Generator with neutral earthed through an impedance.

Since neither positive- nor negative-phase sequence currents may flow in the neutral it follows that

$$V'_{a1} = V'_{a2} = 0,$$

and so the only voltage appearing between N and earth must be of zero-phase sequence, i.e.

$$V_n = V'_a = V'_{a0} = -I_{a0}(3Z_n), \tag{74}$$

$$Z_0 = 3Z_n. \tag{75}$$

The zero-sequence impedance is thus three times the neutral impedance.

The only e.m.f. generated in the a-phase will be of positive-phase sequence E_{a1}. The zero- and negative-phase sequence e.m.f.s E_{a0} and E_{a2} will both be zero, since the generated e.m.f.s are assumed to be balanced, so that

$$\left.\begin{aligned} E_{a1} &= E_a, \\ E_{a2} &= 0, \\ E_{a0} &= 0. \end{aligned}\right\} \tag{76}$$

It is now possible to derive the symmetrical components of the terminal voltage between line and earth for the a-phase. Since the terminal voltage is equal to the generated e.m.f. minus the impedance voltage drop, then

$$V_{a1} = E_{a1} - I_{a1} Z_{G1} \qquad = E_a - I_{a1} Z_{G1}, \tag{77}$$

$$V_{a2} = E_{a2} - I_{a2} Z_{G2} \qquad = - I_{a2} Z_{G2}, \tag{78}$$

$$V_{a0} = E_{a0} - I_{a0} (Z_0' + Z_{G0}) \qquad = - I_{a0}(3Z_n + Z_{G0}), \tag{79}$$

where Z_{G1}, Z_{G2}, and Z_{G0} are the generator positive-, negative-, and zero-sequence impedances respectively.

With no impedance in the neutral and with the star point solidly earthed,

$$3Z_n = Z_0' = 0,$$

and eqn. (79) becomes

$$V_{a0} = - I_{a0} Z_{G0}. \tag{80}$$

If the star point is isolated from earth, Z_n will be infinite.

28. Kirchhoff's first law

Kirchhoff's first law may be stated in the form "The vector sum of all the currents flowing in conductors which meet at a point in a circuit is zero", i.e.

$$\Sigma I = 0.$$

Referring to Fig. 23, where I'_a, I''_a and I'''_a are the currents flowing in three conductors meeting at a point in the a-phase of a three-phase circuit, this means that

$$I'_a + I''_a + I'''_a = 0, \tag{81}$$

and similarly for the b- and c-phases

$$I'_b + I''_b + I'''_b = 0, \tag{82}$$

$$I'_c + I''_c + I'''_c = 0. \tag{83}$$

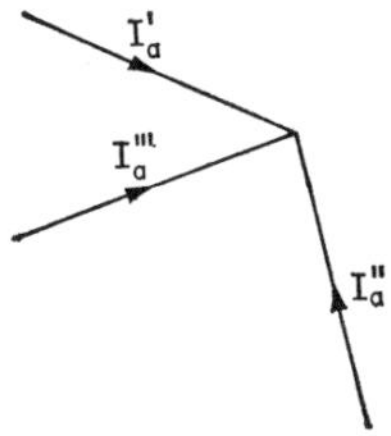

FIG. 23. Kirchhoff's first law.

Expanding eqn. (81) in terms of the symmetrical components of the three currents gives

$$(I'_{a0} + I'_{a1} + I'_{a2}) + (I''_{a0} + I''_{a1} + I''_{a2}) + (I'''_{a0} + I'''_{a1} + I'''_{a2}) = 0,$$

$$(I'_{a0} + I''_{a0} + I'''_{a0}) + (I'_{a1} + I''_{a1} + I'''_{a1}) + (I'_{a2} + I''_{a2} + I'''_{a2}) = 0, \tag{84}$$

and similar expansions of eqns. (82) and (83) give

$$(I'_{a0} + I''_{a0} + I'''_{a0}) + h^2(I'_{a1} + I''_{a1} + I'''_{a1}) + h(I'_{a2} + I''_{a2} + I'''_{a2}) = 0, \tag{85}$$

$$(I'_{a0} + I''_{a0} + I'''_{a0}) + h(I'_{a1} + I''_{a1} + I'''_{a1}) + h^2(I'_{a2} + I''_{a2} + I'''_{a2}) = 0. \tag{86}$$

Adding eqns. (84), (85), and (86) and using the relationship $(1 + h + h^2) = 0$ gives

$$(I'_{a0} + I''_{a0} + I'''_{a0}) = 0. \tag{87}$$

The addition of eqn. (84), h times (85), and h^2 times (86) gives

$$(I'_a + I''_{a1} + I'''_{a1}) = 0, \tag{88}$$

and the addition of eqn. (84), h^2 times (85), and h times (86) gives

$$(I'_{a2} + I''_{a2} + I'''_{a2}) = 0. \tag{89}$$

It is now evident from eqns. (87)–(89) that the separate symmetrical components of the currents meeting at a point in a current obey Kirchhoff's first law in the same way that the total currents do. Thus at any point in a circuit

$$\left.\begin{aligned} \Sigma I_{a0} &= 0, \\ \Sigma I_{a1} &= 0, \\ \Sigma I_{a2} &= 0, \end{aligned}\right\} \tag{90}$$

and similar conditions apply to the b- and c-phases also.

29. Kirchhoff's second law

Kirchhoff's second law may be stated in the form "The vector sum of the e.m.f.s in a closed circuit is equal to the vector sum of the products of the currents and impedances in the branches of that circuit", i.e.

$$\Sigma E = \Sigma IZ.$$

In § 24 it was shown that an e.m.f. of one particular phase sequence produced currents only of the same phase sequence, and the further consideration of Example 9 served to show that in a star-connected load with unbalanced voltages applied to it, the complete circuit may be considered as three separate circuits each of its own separate phase sequence. Each of the individual sequence networks thus obtained obey Kirchhoff's second law, and in general it may be said that for any closed circuit the vector sum of the e.m.f.s of any phase sequence must be equal to the sum of the voltage drops of that same sequence,

$$\left.\begin{aligned} \Sigma E_{a0} &= \Sigma I_{a0} Z_0, \\ \Sigma E_{a1} &= \Sigma I_{a1} Z_1, \\ \Sigma E_{a2} &= \Sigma I_{a2} Z_2. \end{aligned}\right\} \tag{91}$$

Similar identities may, of course, be written for the *b*- and *c*-phase quantities.

30. Star–delta transformations

(a) *Currents*

When dealing with delta-connected impedances, particularly in transformers, or sometimes in generators, it is sometimes useful to be able to convert from line to phase values of symmetrical components and vice versa.

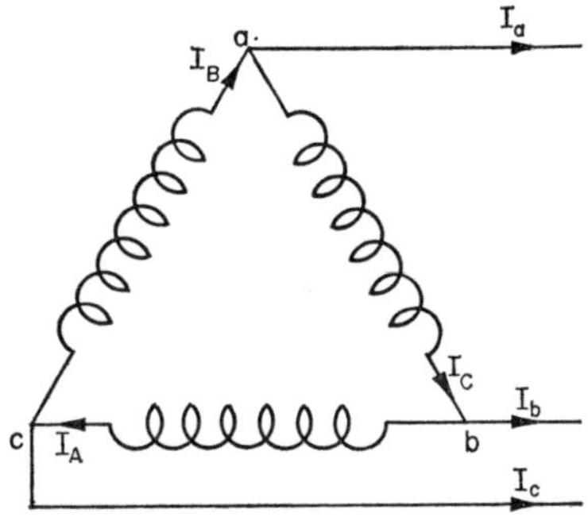

FIG. 24. Line and phase currents in delta-connected impedances.

Let I_a, I_b, and I_c be the line currents and I_A, I_B, and I_C the phase currents as shown in Fig. 24. Having chosen the subscripts for the line currents it is purely arbitrary how the subscripts for the phase values are arranged. Here it has been assumed that the current I_A flows in the impedance opposite line *a* and similarly for I_B and I_C. If other arrangements are chosen the same result will ensue in the final solution.

Applying Kirchhoff's first law at the point *a* in the network for each of the phase sequences in turn we have for the zero-phase sequence components

$$I_{a0} = I_{B0} - I_{C0},$$
$$= 0. \tag{92}$$

For the positive-phase sequence components

$$\begin{aligned} I_{a1} &= I_{B1} - I_{C1}, \\ &= h^2 I_{A1} - h I_{A1}, \\ &= (h^2 - h)\, I_{A1}, \\ &= -j\sqrt{(3)}\, I_{A1}. \end{aligned} \tag{93}$$

Therefore

$$I_{A1} = \frac{j}{\sqrt{3}} I_{a1}, \tag{94}$$

and for the negative-phase sequence components

$$\begin{aligned} I_{a2} &= I_{B2} - I_{C2}, \\ &= h I_{A2} - h^2 I_{A2}, \\ &= (h - h^2)\, I_{A2}, \\ &= j\sqrt{(3)}\, I_{A2}. \end{aligned} \tag{95}$$

Therefore

$$I_{A2} = -\frac{j}{\sqrt{3}} I_{a2}. \tag{96}$$

The same relationships can be shown to apply between components of the *b*- and *c*-phases and are shown in the vector diagrams of Fig. 25 for the positive- and negative-phase sequence currents.

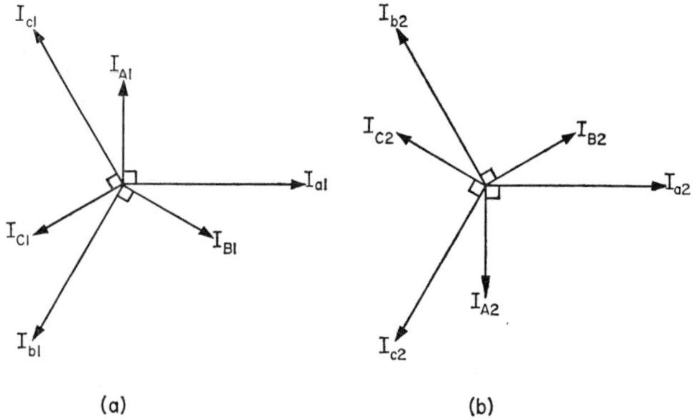

FIG. 25. Positive- and negative-phase sequence current vectors.

(b) *Voltages*

In Fig. 26 the phase voltages are denoted by V_a, V_b, and V_c and the corresponding line voltages by V_A, V_B, and V_C.

Now since

$$V_A = V_c - V_b$$

and since Kirchhoff's second law applies to the symmetrical components we may write

$$\begin{aligned} V_{A0} &= V_{c0} - V_{b0}, \\ &= 0. \end{aligned} \tag{97}$$

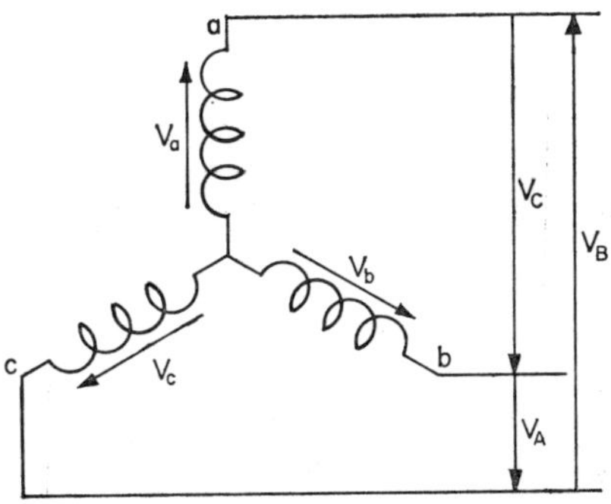

FIG. 26. Line and phase voltages in star-connected impedances.

Also we may write

$$\begin{aligned} V_{A0} &= V_{c1} - V_{b1}, \\ &= hV_{a1} - h^2V_{a1}, \\ &= (h - h^2)\, V_{a1}, \\ &= j\,\sqrt{(3)}\, V_{a1}, \end{aligned} \tag{98}$$

$$V_{a1} = -\frac{j}{\sqrt{3}}\, V_{A1}, \tag{99}$$

and for the negative-phase sequence components

$$V_{A2} = V_{c2} - V_{b2},$$

$$= h^2 V_{a2} - h V_{a2},$$

$$= (h^2 - h)\, V_{a2},$$

$$= -j\sqrt{(3)}\, V_{a2}, \tag{100}$$

$$V_{a2} = \frac{j}{\sqrt{3}}\, V_{A2}. \tag{101}$$

The vector relationships between the line and phase components are shown in Fig. 27.

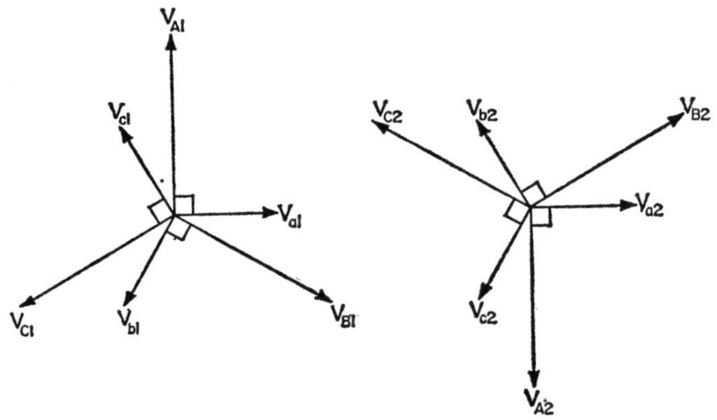

FIG. 27. Positive- and negative-phase sequence voltage vectors.

31. Star–delta transformers

In Fig. 28 is shown a star–delta transformer having a turns ratio of n. In the following theory it will be shown that in addition to a change in the magnitude of the voltages and currents between primary and secondary windings there is also a phase shift which is different for both the positive- and negative-phase sequence components of voltage and current.

Since the primary winding is connected in star and the secondary winding in delta the line voltage ratio N will be given by

$$N = \sqrt{(3)}\, n.$$

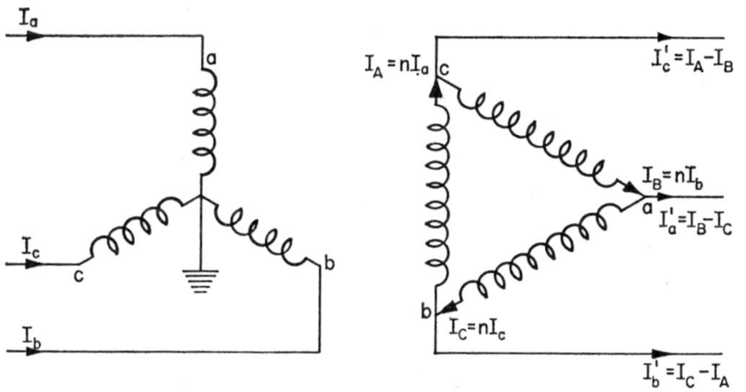

FIG. 28. Currents in a star–delta transformer.

(a) *Currents*

Let I_a, I_b, and I_c be the line currents in the star-connected winding and I'_a, I'_b, and I'_c the line currents in the delta winding. The phase currents in the delta are represented by I_A, I_B, and I_C. Then

$$I_A = nI_a,$$

$$I_B = nI_b,$$

$$I_C = nI_c,$$

and also

$$\left.\begin{aligned} I'_a &= I_B - I_C = n(I_b - I_c) = \frac{N}{\sqrt{3}}(I_b - I_c), \\ I'_b &= I_C - I_A = n(I_c - I_a) = \frac{N}{\sqrt{3}}(I_c - I_a), \\ I'_{ac} &= I_A - I_B = n(I_a - I_b) = \frac{N}{\sqrt{3}}(I_a - I_b). \end{aligned}\right\} \quad (102)$$

Considering now only the positive-phase sequence components of currents we have from eqn. (102)

$$\left.\begin{aligned} I'_{a1} &= \frac{N}{\sqrt{3}}(I_{b1} - I_{c1}) = \frac{N}{\sqrt{3}}(h^2 - h)\, I_{a1} = -jNI_{a1}, \\ I'_{b1} &= \frac{N}{\sqrt{3}}(I_{c1} - I_{a1}) = \frac{N}{\sqrt{3}}(h^2 - h)\, I_{b1} = -jNI_{b1}, \\ I'_{c1} &= \frac{N}{\sqrt{3}}(I_{a1} - I_{b1}) = \frac{N}{\sqrt{3}}(h^2 - h)\, I_{c1} = -jNI_{c1}. \end{aligned}\right\} \tag{103}$$

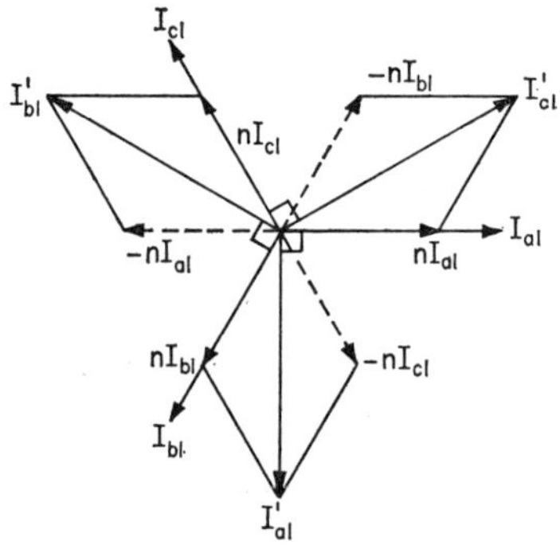

FIG. 29. Positive-phase sequence currents.

Thus, with the current relationships which have been chosen between the star and the delta, there is a phase shift of $-90°$ in addition to the change in magnitude of the currents of the positive-phase sequence. This is shown in the vector diagram of Fig. 29.

Applying eqns. (102) to the negative-phase sequence components of currents we have

$$\left.\begin{aligned} I_{a2}' &= \frac{N}{\sqrt{3}}(I_{b2} - I_{c2}) = \frac{N}{\sqrt{3}}(h - h^2)\, I_{a2} = jNI_{a2}, \\ I_{b2}' &= \frac{N}{\sqrt{3}}(I_{c2} - I_{a2}) = \frac{N}{\sqrt{3}}(h - h^2)\, I_{b2} = jNI_{b2}, \\ I_{c2}' &= \frac{N}{\sqrt{3}}(I_{a2} - I_{b2}) = \frac{N}{\sqrt{3}}(h - h^2)\, I_{c2} = jNI_{c2}. \end{aligned}\right\} \quad (104)$$

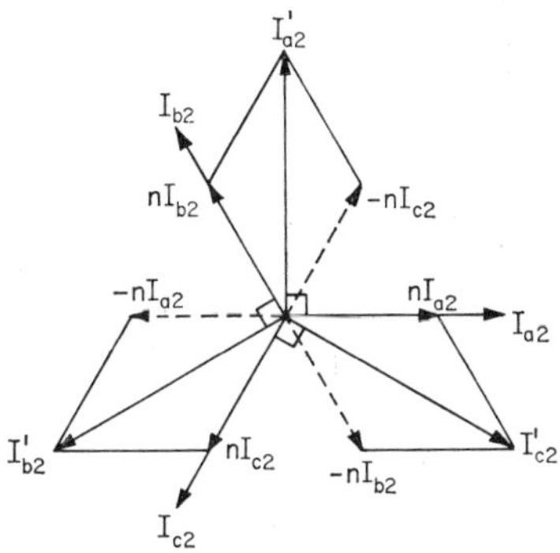

FIG. 30. Negative-phase sequence currents.

The phase shift between the corresponding negative-phase sequence currents in the star and delta windings is now seen to be $+90°$, and is shown in the vector diagram of Fig. 30. For the zero-phase sequence components of current we have

$$I_{a0}' = \frac{N}{\sqrt{3}}(I_{b0} - I_{c0}) = 0, \quad (105)$$

and, similarly, $I_{b}'{}_0$ and $I_{c}'{}_0$ are both zero. Whilst zero-phase sequence components of current may flow in the lines of the star-connected winding, there can be no zero-phase sequence components in the lines of the delta winding. They will, however, flow round the delta if present in the star.

(b) *Voltages*

In the ideal transformer of Fig. 31, the line voltages of the primary and secondary windings are represented by E_a, E_b, E_c and E'_a, E'_b, E'_c respectively, and the phase voltages of the primary winding by E_A, E_B, E_C.

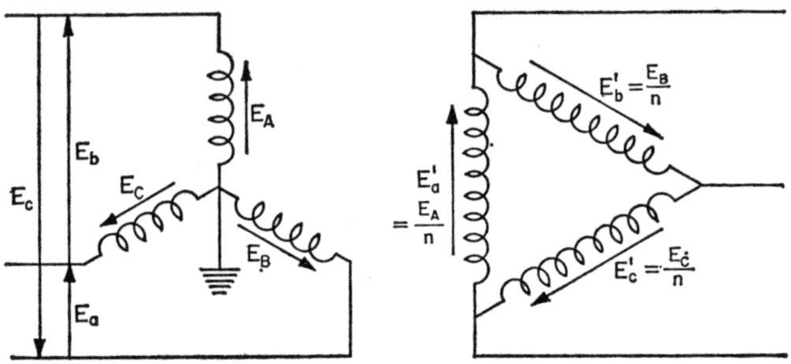

FIG. 31. Voltages in a star–delta transformer.

Now for the star-connected winding the following relationships hold:

$$\left.\begin{aligned} E_a &= E_C - E_B = n(E'_c - E'_b) = \frac{N}{\sqrt{3}}(E'_c - E'_b), \\ E_b &= E_A - E_C = n(E'_a - E'_c) = \frac{N}{\sqrt{3}}(E'_a - E'_c), \\ E_c &= E_B - E_A = n(E'_b - E'_a) = \frac{N}{\sqrt{3}}(E'_b - E'_a). \end{aligned}\right\} \quad (106)$$

Writing the above expressions for the positive-phase sequence components gives:

$$\left.\begin{aligned}
E_{a1} &= \frac{N}{\sqrt{3}}(E'_{c1} - E'_{b1}) = \frac{N}{\sqrt{3}}(h - h^2)\, E'_{a1} = jNE'_{a1},\\
E_{b1} &= \frac{N}{\sqrt{3}}(E'_{a1} - E'_{c1}) = \frac{N}{\sqrt{3}}(h - h^2)\, E'_{b1} = jNE'_{b1},\\
E_{c1} &= \frac{N}{\sqrt{3}}(E'_{b1} - E'_{a1}) = \frac{N}{\sqrt{3}}(h - h^2)\, E'_{c1} = jNE'_{c1},
\end{aligned}\right\} \tag{107}$$

from which

$$\left.\begin{aligned}
E'_{a1} &= -\frac{j}{N}\, E_{a1},\\
E'_{b1} &= -\frac{j}{N}\, E_{b1},\\
E'_{c1} &= -\frac{j}{N}\, E_{c1}.
\end{aligned}\right\} \tag{108}$$

As with the positive-phase sequence components of the line currents it is seen that there is also a phase shift of $-90°$ in the corresponding components of the line voltages. The currents and voltages must undergo the same phase shift in order that the power and reactive kVA are equal for both primary and secondary sides of the ideal transformer.

Expressing now the negative-phase sequence components in eqns. (106) gives:

$$\left.\begin{aligned}
E_{a2} &= \frac{N}{\sqrt{3}}(E'_{c2} - E_{b2}) = \frac{N}{\sqrt{3}}(h^2 - h)\, E'_{a2} = -jNE'_{a2},\\
E_{b2} &= \frac{N}{\sqrt{3}}(E'_{a2} - E_{c2}) = \frac{N}{\sqrt{3}}(h^2 - h)\, E'_{b2} = -jNE'_{b2},\\
E_{c2} &= \frac{N}{\sqrt{3}}(E'_{b2} - E'_{a2}) = \frac{N}{\sqrt{3}}(h^2 - h)\, E'_{c2} = -jNE'_{c2},
\end{aligned}\right\} \tag{109}$$

which gives

$$\left.\begin{aligned} E_{a2}' &= \frac{j}{N}\,E_{a2}, \\ E_{b2}' &= \frac{j}{N}\,E_{b2}, \\ E_{c2}' &= \frac{j}{N}\,E_{c2}. \end{aligned}\right\} \qquad (110)$$

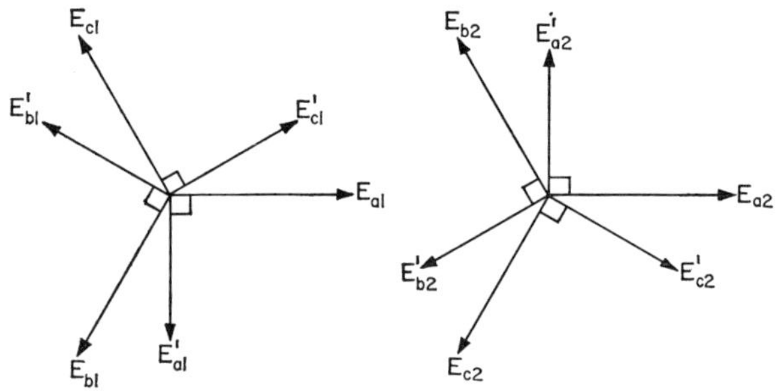

FIG. 32. Positive- and negative-phase sequence voltages.

Again the phase shift between the negative-phase sequence components in going from primary to secondary is $+90°$ as in the case of the currents.

The vector relationships between the positive- and negative-phase sequence components of the line voltages is shown in Fig. 32.

So far as the zero-phase sequence components of voltage are concerned, whilst they may appear in the phases, they will be absent between lines and in the case of the secondary winding will cause current to circulate round the delta.

32. Effects of unbalanced loading on synchronous machines

When a three-phase synchronous machine is supplying an unbalanced load the armature currents may be resolved into their three symmetrical components.

The positive-phase sequence components of current will set up in the air gap an armature reaction m.m.f. which rotates at synchronous speed relative to the armature and which is stationary relative to the main field system of the machine. Its effect upon the main field is one of armature reaction.

Due to the electrical displacement of 120° between the three-phase windings, the resultant m.m.f. produced by the zero-phase sequence current components will be zero and consequently it will have no effect.

Negative-phase sequence currents, however, will produce in the air gap an m.m.f. rotating at synchronous speed relative to the armature in the direction opposite to that produced by the positive-phase sequence components, and relative to the main field this m.m.f. will rotate at twice synchronous speed. This will cause currents of double frequency to be induced in the main field system. In the case of an unlaminated rotating field machine these induced currents, being of high frequency, will tend to flow along the periphery of the rotor along the wedges and through the damper winding where the reactance is lowest. If allowed to become excessive they will cause serious overheating of parts of the rotor body, particularly at the two ends of the rotor where the current density will be greatest. It is thus necessary to limit the permissible value of negative-phase sequence current due to the unbalanced loading of a generator. The maximum permissible value of negative-phase sequence current which may be tolerated depends upon how highly rated a machine is. Typical continuous values may vary between about 10% and 20% of the rated current in the machine provided that the rated current is not exceeded in any one phase. Higher values may be possible for a shorter period of time which may be determined by the heating effect of these currents. Usually a permissible value of $I_2^2 t$ is specified by the machine designer for a particular type of machine, where $I_2 =$ negative phase sequence current in per unit of rated stator current and $t =$ time in seconds.

Typical values of $I_2^2 t$ may range between about 4 and 30 depending upon the particular type of machine.

33. Power due to unbalanced three-phase loads

The instantaneous power at a point in a three-phase system is the sum of the power in each of the three phases and is given in terms of the instantaneous values of the three-phase voltages and currents by

$$p = v_a i_a + v_b i_b + v_c i_c. \tag{111}$$

Replacing the instantaneous values of phase voltage and current by their instantaneous values of symmetrical components gives

$$\begin{aligned} p &= (v_{a0} + v_{a1} + v_{a2})\,(i_{a0} + i_{a1} + i_{a2}) \\ &\quad + (v_{b0} + v_{b1} + v_{b2})\,(i_{b0} + i_{b1} + i_{b2}) \\ &\quad + (v_{c0} + v_{c1} + v_{c2})\,(i_{c0} + i_{c1} + i_{c2}), \\ &= 3\,v_{a0}\,i_{a1} + (v_{a1}\,i_{a1} + v_{b1}\,i_{b1} + v_{c1}\,i_{c1}) \\ &\quad + (v_{a2}\,i_{a2} + v_{b2}\,i_{b2} + v_{c2}\,i_{c2}) + (v_{a1}\,i_{a2} + v_{b1}\,i_{b2} + v_{c1}\,i_{c2}) \\ &\quad + (v_{a2}\,i_{a1} + v_{b2}\,i_{b1} + v_{c2}\,i_{c1}); \end{aligned} \tag{112}$$

since $v_{a0} = v_{b0} = v_{c0}$,

$$i_{a0} = i_{b0} = i_{c0},$$

$$i_{a1} + i_{b1} + i_{c1} = 0,$$

and $i_{a2} + i_{b2} + i_{b2} = 0.$

The instantaneous phase values of current and voltage may be represented in terms of the r.m.s. values as follows:

$$\left.\begin{aligned}
v_{a0} &= v_{b0} = v_{c0} = \sqrt{(2)}\,|\,V_{a0}\,|\sin(\omega t + \theta_0),\\
i_{a0} &= i_{b0} = i_{c0} = \sqrt{(2)}\,|\,I_{a0}\,|\sin(\omega t + \theta_0 + \phi_0),\\
v_{a1} &= \sqrt{(2)}\,|\,V_{a1}\,|\sin(\omega t + \theta_1),\\
v_{b1} &= \sqrt{(2)}\,|\,V_{a1}\,|\sin\left(\omega t + \theta_1 + \frac{4\pi}{3}\right),\\
v_{c1} &= \sqrt{(2)}\,|\,V_{a1}\,|\sin\left(\omega t + \theta_1 + \frac{2\pi}{3}\right),\\
i_{a1} &= \sqrt{(2)}\,|\,I_{a1}\,|\sin(\omega t + \theta_1 + \phi_1),\\
i_{b1} &= \sqrt{(2)}\,|\,I_{a1}\,|\sin\left(\omega t + \theta_1 + \phi_1 + \frac{4\pi}{3}\right),\\
i_{c1} &= \sqrt{(2)}\,|\,I_{a1}\,|\sin\left(\omega t + \theta_1 + \phi_1 + \frac{2\pi}{3}\right),\\
v_{a2} &= \sqrt{(2)}\,|\,V_{a2}\,|\sin(\omega t + \theta_2),\\
v_{b2} &= \sqrt{(2)}\,|\,V_{a2}\,|\sin\left(\omega t + \theta_2 + \frac{2\pi}{3}\right),\\
v_{c2} &= \sqrt{(2)}\,|\,V_{a2}\,|\sin\left(\omega t + \theta_2 + \frac{4\pi}{3}\right),\\
i_{a2} &= \sqrt{(2)}\,|\,I_{a2}\,|\sin(\omega t + \theta_2 + \phi_2),\\
i_{b2} &= \sqrt{(2)}\,|\,I_{a2}\,|\sin\left(\omega t + \theta_2 + \phi_2 + \frac{2\pi}{3}\right),\\
i_{c2} &= \sqrt{(2)}\,|\,I_{a2}\,|\sin\left(\omega t + \theta_2 + \phi_2 + \frac{4\pi}{3}\right),
\end{aligned}\right\} \quad (113)$$

where $|\,V_{a1}\,|$, $|\,I_{a1}\,|$, etc., are the scalar r.m.s. values of voltage and current.

Substituting the instantaneous values given by eqns. (113) in eqn. (114) and making use of the relationship $\sin A \sin B = \frac{1}{2}[\cos(A - B) - \cos(A + B)]$ gives

$$\left.\begin{aligned} p = 3\,[\,|V_{a0}|\,|I_{a0}|\cos\phi_0 + |V_{a1}|\,|I_{a1}|\cos\phi_1 \\ + |V_{a2}|\,|I_{a2}|\cos\phi_2] \\ -3\,|V_{a0}|\,|I_{a0}|\cos(2\omega t + 2\theta_0 + \phi_0) \\ -3\,|V_{a1}|\,|I_{a2}|\cos(2\omega t + \theta_1 + \theta_2 + \phi_2) \\ -3\,|V_{a2}|\,|I_{a1}|\cos(2\omega t + \theta_1 + \theta_2 + \phi_1). \end{aligned}\right\} \quad (114)$$

The first three terms are constant whilst the last three whose average value will be zero, pulsate at double frequency. The average value of three-phase power will thus be given by

$$P = 3\,[\,|V_{a0}|\,|I_{a0}|\cos\phi_0 + |V_{a1}|\,|I_{a1}|\cos\phi_1 + |V_{a2}|\,|I_{a2}|\cos\phi_2]. \quad (115)$$

In terms of the r.m.s. unbalanced phase voltages and currents the power is also given by

$$P = |V_a|\,|I_a|\cos\phi_a + |V_b|\,|I_b|\cos\phi_b + |V_c|\,|I_c|\cos\phi_c. \quad (116)$$

Considering for a moment the single-phase case, the instantaneous values of voltage and current are given by

$$\begin{aligned} v &= \sqrt{2}\,|V|\sin(\omega t + \theta), \\ i &= \sqrt{2}\,|I|\sin(\omega t + \theta + \phi), \end{aligned} \quad (117)$$

where $|V|$ and $|I|$ are the scalar r.m.s. values of voltage and current. The real power is given by

$$P = |V|\,|I|\cos\phi, \quad (118)$$

and the reactive volt amperes by

$$Q = |V|\,|I|\sin\phi. \quad (119)$$

The real and reactive volt amperes may then be expressed by S,

where $$S = P + jQ,$$
$$= |V||I|(\cos\phi + j\sin\phi). \tag{120}$$

Also the voltage and current vectors are given by

$$V = |V|e^{j\theta},$$
$$I = |I|e^{j(\theta+\varphi)}, \tag{121}$$

where V and I are both vector quantities.

Now eqn. (120) may be written as

$$S = |V|e^{-j\theta}|I|e^{j(\theta+\varphi)},$$
$$= V^*I, \tag{122}$$

where V^* is the conjugate of V. From eqn. (122), P is the real part of V^*I and Q is the imaginary part.

This may now be extended to the three-phase case where the total active and reactive power is given by

$$S = V_a^* I_a + V_b^* I_b + V_c^* I_c \tag{123}$$

or, in terms of the symmetrical components, we have

$$S = 3(V_{a0}^* I_{a0} + V_{a1}^* I_{a1} + V_{a2}^* I_{a2}), \tag{124}$$

where, as in the single-phase case, the active volt amperes P is the real part of S and the reactive volt amperes Q is the imaginary part of S.

Exercises

1. An unbalanced star-connected load consists of a capacitor of reactance $X_A = -(j10\sqrt{3})$ ohms, and two resistors $R_B = 60$ ohms and $R_C = 30$ ohms, as shown in Fig. 33, The power is measured by the two-wattmeter method with wattmeters W_1 and W_2. The load is connected to a symmetrical 3-phase supply ABC with line voltages $V_{AB} = 300$ V, $V_{BC} = h^2 300$ V and $V_{CA} = h300$ V, where $h = \frac{1}{2}(-1 + j\sqrt{3})$. Calculate the readings of the wattmeters W_1 and W_2 and the power in each arm of the load. [I.E.E., November 1962]

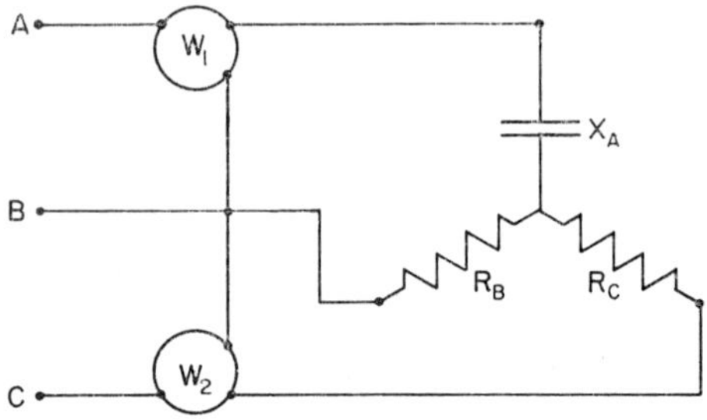

FIG. 33. Exercise 1.

2. A 132 kV 3-phase substation supplies a 3-phase balanced load of 40 MW at 0·8 lagging power factor and a single-phase railway load of 10 MW at 0·9 lagging power factor connected between two lines. The substation is fed through step-up transformers from a 60 MW, 11 kV cylindrical rotor generator. Determine the positive- and negative-sequence components of current in the generator. Neglect losses and voltage drops.

Explain why the single-phase load may give rise to serious overheating in the generator. [I.E.E., November 1961]

CHAPTER 4

The Measurement of Symmetrical Components

SO FAR, the mathematical analysis of unbalanced three-phase vectors into their symmetrical components has been considered. It is now necessary to consider means by which these components may be measured by the use of instruments and also to examine various filter circuits which may be employed in protection equipment for the suppression of various phase sequence components.

In general the methods of measurement are quite straightforward, using suitable connections of instrument transformers to give the required quantity which may then be measured in the usual way.

34. Zero-sequence voltage

It has been shown in Chapter 2 that the zero-phase sequence component of voltage is equal to one-third of the residue of the phase voltages, i.e.

$$V_{a0} = \tfrac{1}{3}(V_a + V_b + V_c).$$

Use of this relationship is made in the measurement of zero-sequence voltage. Three single-phase voltage transformers are connected as shown in Fig. 34 with their primary windings connected in star with their star point connected to the neutral and the opposite ends of the windings connected to the three lines. The secondary windings are connected in open delta so that the voltage across them will be proportional to the vector sum of the three-phase voltages. A voltmeter V, shown in the circuit, is used to measure this voltage and if suitably calibrated may be used to measure the zero-sequence voltage.

A point worthy of note here is that it is important that the voltage transformers must be quite separate and magnetically isolated, otherwise the true value of residual voltage will not be measured.

A second method which may be used is shown in Fig. 35. Three accurately balanced impedances Z are connected in star and to the three lines. The star point is then connected through a voltmeter

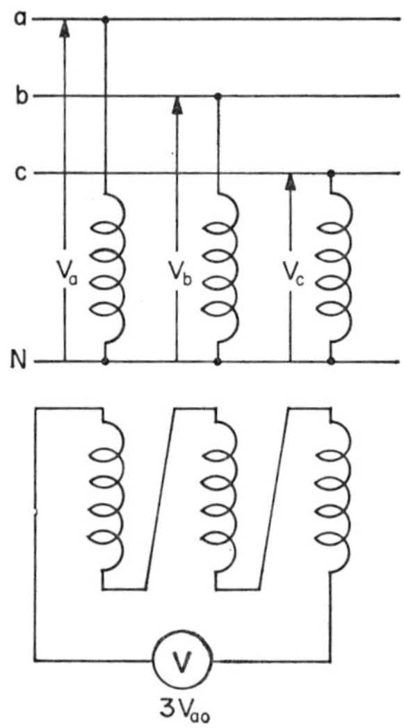

FIG. 34. Measurement of zero-phase sequence voltage—Method 1.

to neutral. Here it is necessary for the voltmeter to be calibrated to take into account the three parallel impedances connected in series with it.

35. Zero-sequence current

As for voltage, the zero-phase sequence component of current is one-third of the residue of the three-phase current (Fig. 36), where

$$I_{a0} = \tfrac{1}{3}(I_a + I_b + I_c).$$

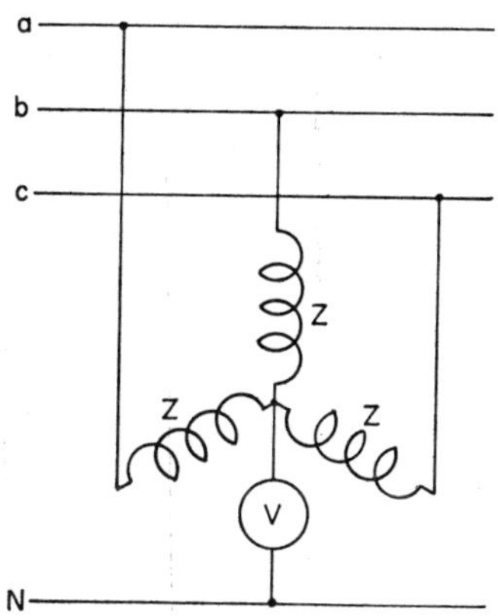

FIG. 35. Measurement of zero-phase sequence voltage—Method 2.

The method of measurement involves the use of three current transformers whose secondary windings are connected in parallel. The ammeter *A*, connected across the three secondary windings, will carry a current corresponding to the vector sum of the three-line currents and hence may be calibrated to read the zero-phase sequence current.

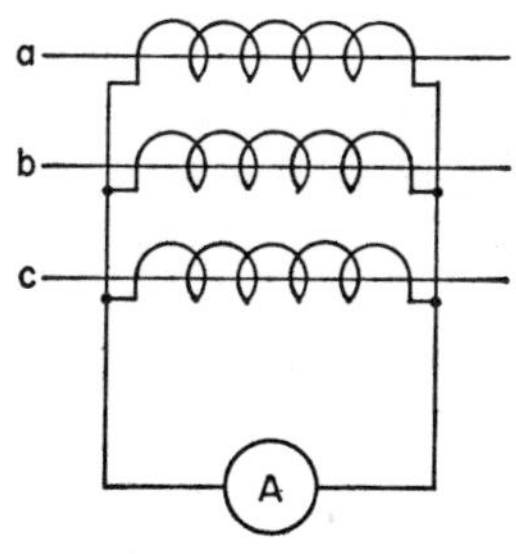

FIG. 36. Measurement of zero-phase sequence current.

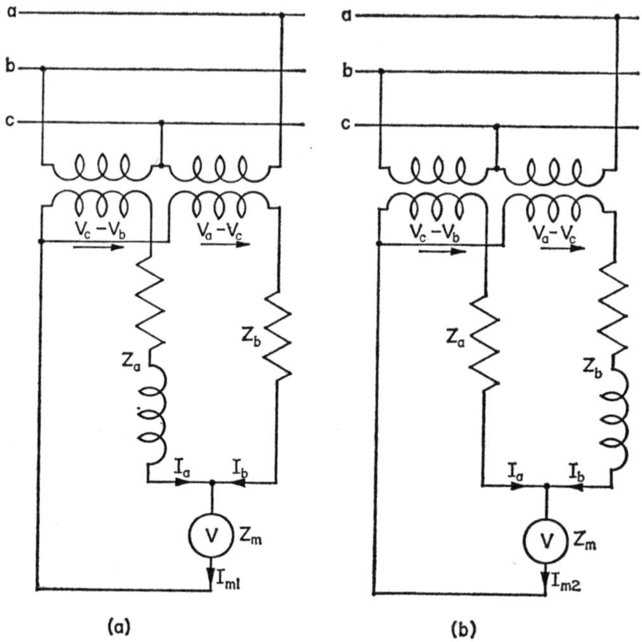

FIG. 37. Measurement of positive- and negative-phase sequence voltages.

36. Positive- and negative-sequence voltage

The networks shown in Fig. 37a, b may be used respectively for the measurement of positive- and negative-phase sequence voltages. These networks make use of the measurement of line-to-line voltages and so eliminate any zero-phase sequence voltages. It will also be noted that the connections to the secondary winding of one voltage transformer are reversed. Let the meter have an impedance Z_m and a current flowing through it of I_m, where from the diagram

$$I_m = I_a + I_b.$$

Then, applying Kirchhoff's laws,

$$\left.\begin{aligned} V_a - V_c &= I_b Z_b + I_m Z_m, \\ V_c - V_b &= I_a Z_a + I_m Z_m. \end{aligned}\right\}$$

Therefore
$$\left.\begin{aligned} Z_a (V_a - V_c) &= I_b Z_a Z_b + I_m Z_a Z_m, \\ Z_b (V_c - V_b) &= I_a Z_a Z_b + I_m Z_b Z_m. \end{aligned}\right\}$$

Adding the above two equations, and since $(I_a + I_b) = Z_m$, gives

$$Z_a (V_a - V_c) + Z_b (V_c - V_b) = I_m[Z_a Z_b + Z_m(Z_a + Z_b)]$$

from which

$$I_m = \frac{Z_a (V_a - V_c) + Z_b (V_c - V_b)}{Z_a Z_b + Z_m (Z_a + Z_b)}.$$

Expressing the voltages in terms of their symmetrical components gives

$$I_m = \frac{Z_a[V_{a1}(1-h) + V_{a2}(1-h^2)] + Z_b[V_{a1}(h-h^2) + V_{a2}(h^2-h)]}{Z_a Z_b + Z_m (Z_a + Z_b)},$$

$$= \frac{V_{a1}[(1-h)(Z_a + hZ_b)] + V_{a2}[(1-h^2)(Z_a + h^2 Z_b)]}{Z_a Z_b + Z_m (Z_a + Z_b)}. \quad (125)$$

The term in V_{a2} will disappear if

$$Z_a + h^2 Z_b = 0 \quad \text{or } Z_a = -h^2 Z_b.$$

If now, in Fig. 37a, Z_b is made purely resistive and equal to R, we have

$$Z_b = R,$$

$$Z_a = -h^2 R = R(\tfrac{1}{2} + \tfrac{1}{2} j\sqrt{3}),$$

where Z_a is an inductive impedance having a power factor angle of $+60°$.

With the above values of Z_a and Z_b the instrument may be used to measure positive phase sequence voltage and the current I_{m1} passing through it will be given by

$$I_{m1} = \frac{V_{a1}\,[(1-h)(-h^2Z_b + hZ_b)]}{-h^2Z_b^2 + Z_m(-h^2Z_b + Z_b)},$$

$$= \frac{V_{a1}\,h(1-h)^2}{-h^2Z_b + Z_m(1-h^2)},$$

$$= \frac{3V_{a1}}{Z_b + Z_m(1-h)}. \tag{126}$$

The current passing through the instrument is thus seen to be dependent upon the positive phase sequence voltage V_{a1}, and Z_b and Z_m which are constants of the network.

Returning again to eqn. (125) it may be seen that the term in V_{a2} will be zero if

$$Z_a + hZ_b = 0 \quad \text{or} \quad Z_b = -h^2Z_a.$$

If, in Fig. 37b, Z_a is made equal to R, then

$$Z_a = R,$$

$$Z_b = -h^2R = R(\tfrac{1}{2} + j\tfrac{1}{2}\sqrt{3}).$$

Using these values of Z_a and Z_b the instruments can now be used to measure the negative-phase sequence voltage when the current I_{m2} passing through it will be given by

$$I_{m2} = \frac{V_{a2}[(1-h^2)(-hZ_b + h^2Z_b)]}{-hZ_b^2 + Z_m(-hZ_b + Z_b)},$$

$$= \frac{V_{a2}\,h\,(h^2 + h - 2)}{-hZ_b + Z_m(1-h)},$$

$$= \frac{3V_{a2}}{Z_b + Z_m(1-h^2)} \tag{127}$$

37. Simultaneous measurement of positive- and negative-sequence voltage

The two networks of Fig. 37a, b may be combined as in Fig. 38 to measure simultaneously both the positive- and negative-phase sequence voltages. Here Z_b is made purely resistive, whereas Z_a is partly resistive and partly inductive, such that

$$Z_a = -h^2 Z_b.$$

Thus if $$Z_b = R,$$

then $$Z_a = -h^2 R = R(\tfrac{1}{2} + j\tfrac{1}{2}\sqrt{3}).$$

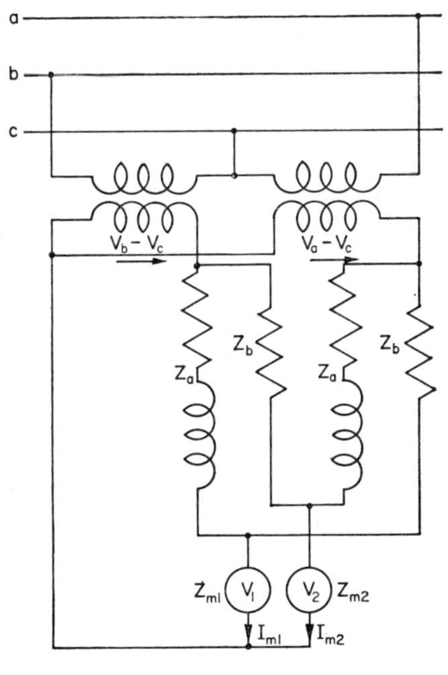

FIG. 38. Simultaneous measurement of positive- and negative-phase sequence voltages.

Let the meter impedances be Z_{m1} and Z_{m2} and their respective currents I_{m1} and I_{m2}. From a comparison of the networks shown in Figs. 37 and 38 and inspection of eqns. (126) and (127) it is easily seen that in Fig. 38 the meter currents are given by

$$I_{m1} = \frac{3V_{a1}}{Z_b + Z_{m1}(1-h)}, \tag{128}$$

and

$$I_{m2} = \frac{3V_{a2}}{Z_a + Z_{m2}(1-h^2)}. \tag{129}$$

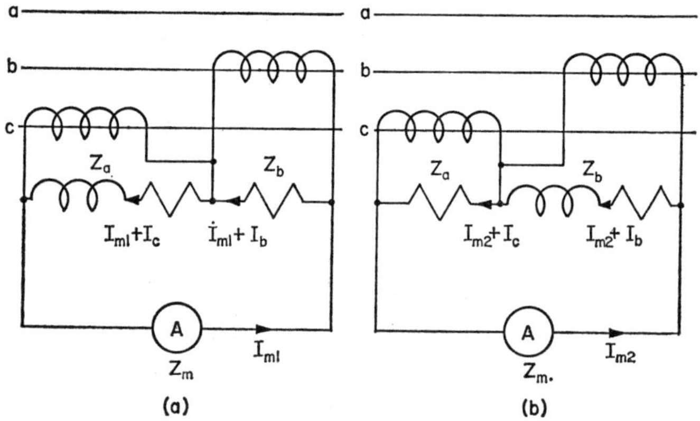

FIG. 39. Measurement of positive- and negative-phase sequence currents without residue.

38. Positive- and negative-sequence current without residue

The circuits shown in Fig. 39 may be used for the measurement of positive- and negative-phase sequence currents respectively provided that zero-phase sequence currents are absent. Across the secondary windings of the two current transformers are connected the impedances Z_a and Z_b. Let the meter current be I_m and its impedance Z_m, then we have

$$I_m Z_m = -(I_m + I_c)Z_a - (I_m + I_b)Z_b.$$

Therefore $I_m (Z_a + Z_b + Z_m) = - I_c Z_a - I_b Z_b.$

Therefore $$I_m = \frac{- I_c Z_a - I_b Z_b}{Z_a + Z_b + Z_m},$$

and if no zero-phase sequence components of current are present we have

$$I_m = \frac{-(hI_{a1} + h^2 I_{a2})Z_a - (h^2 I_{a1} + hI_{a2})Z_b}{Z_a + Z_b + Z_m},$$

$$= \frac{- I_{a1} (hZ_a + h^2 Z_b) - I_{a2}(h^2 Z_a + hZ_b)}{Z_a + Z_b + Z_m}.$$

If in Fig. 39a

$$Z_b = R \quad \text{and} \quad Z_a = R(\tfrac{1}{2} + j\tfrac{1}{2}\sqrt{3}),$$

i.e. $$Z_a = - h^2 Z_b,$$

then $$I_{m1} = \frac{I_{a1} Z_b (1 - h^2)}{(1 - h^2) Z_b + Z_m},$$

$$= \frac{I_{a1}}{1 + \dfrac{Z_m}{Z_b} [(1 - h)/3]}. \tag{130}$$

Hence the meter may be used to measure the positive-phase sequence component of the current since Z_b and Z_m are constants of the network.

If now in Fig. 39b we make

$$Z_a = R,$$

$$Z_b = R(\tfrac{1}{2} + j\tfrac{1}{2}\sqrt{3}),$$

i.e. $$Z_b = - h^2 Z_a,$$

then from eqn. (129) we obtain the result

$$I_{m2} = \frac{I_{a2} Z_b (1 - h)}{(1 - h) Z_b + Z_m},$$

$$= \frac{I_{a2}}{1 + \dfrac{Z_m}{Z_b} [(1 - h^2)/3]}, \tag{131}$$

from which it is seen that the meter may now be used to measure the negative-phase sequence component of current, Z_b and Z_m again being constants of the network.

39. Simultaneous measurement of positive- and negative-sequence currents without residue

An elaborated form of the previous network is shown as a bridge network in Fig. 40, where, provided that zero-sequence components of current are absent, the positive- and negative-phase

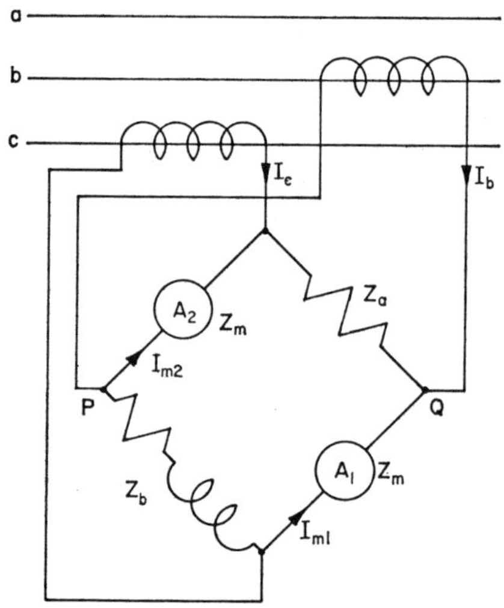

FIG. 40. Simultaneous measurement of positive- and negative-phase sequence currents without residue.

sequence components can be measured simultaneously. The two ammeters A_1 and A_2 in the bridge have equal impedances Z_m. Equating the voltage drops between P and Q through A_1 and Z_b, and A_2 and Z_a, in terms of I_{m1} gives

$$I_{m1}Z_m + (I_c + I_{m1})Z_b = -(I_b + I_c + I_{m1})Z_m - (I_b + I_{m1})Z_a.$$

Therefore

$$I_{m1}(Z_a + Z_b + 2Z_m) = -(Z_a + Z_m)I_b - (Z_b + Z_m)I_c.$$

Therefore

$$I_{m1} = \frac{-(Z_a + Z_m)I_b - (Z_b + Z_m)I_c}{Z_a + Z_b + 2Z_m},$$

and, since zero-phase sequence components are absent, this gives

$$I_{m1} = \frac{-(Z_a + Z_m)(h^2I_{a1} + hI_{a2}) - (Z_b + Z_m)(hI_{a1} + h^2I_{a2})}{Z_a + Z_b + 2Z_m},$$

$$= \frac{-[h^2(Z_a + Z_m) + h(Z_b + Z_m)]I_{a1} - [h(Z_a + Z_m) + h^2(Z_b + Z_m)]I_{a2}}{Z_a + Z_b + 2Z_m}. \quad (132)$$

Equating now the voltage drops between P and Q in terms of I_{m2} gives

$$I_{m2}Z_m + (I_c + I_{m2})Z_a = -(I_b + I_{m2})Z_b - (I_b + I_c + I_{m2})Z_m.$$

Therefore

$$I_{m2}(Z_a + Z_b + 2Z_m) = -(Z_b + Z_m)I_b - (Z_a + Z_m)I_c.$$

Therefore

$$I_{m2} = \frac{-(Z_b + Z_m)I_b - (Z_a + Z_m)I_c}{Z_a + Z_b + 2Z_m},$$

$$= \frac{-(Z_b + Z_m)(h^2I_{a1} + hI_{a2}) - (Z_a + Z_m)(hI_{a1} + h^2I_{a2})}{Z_a + Z_b + 2Z_m},$$

$$= \frac{-[h(Z_a + Z_m) + h^2(Z_b + Z_m)]I_{a1} - [h^2(Z_a + Z_m) + h(Z_b + Z_m)]I_{a2}}{Z_a + Z_b + 2Z_m}, \quad (133)$$

Inspection of eqns. (145) and (146) shows that for I_{m1} to represent the positive-phase sequence current and I_{m2} the negative-phase sequence component,

$$h(Z_a + Z_m) + h^2(Z_b + Z_m) = 0$$

or

$$(Z_b + Z_m) = -h^2(Z_a + Z_m). \tag{134}$$

Using this relationship gives

$$I_{m1} = \frac{(Z_a + Z_m)(1 - h^2)}{(Z_a + Z_m)(1 - h^2)} I_{a1} = I_{a1}, \tag{135}$$

$$I_{m2} = \frac{(Z_a + Z_m)(1 - h^2)}{(Z_a + Z_m)(1 - h^2)} I_{a2} = I_{a2}. \tag{136}$$

Thus it is seen that the ammeters A_1 and A_2 will read, respectively, the positive- and negative-phase sequence currents.

40. Elimination of zero-sequence component

In the two preceding sections, dealing with the measurement of positive- and negative-phase sequence components of current, it has been assumed that the zero-phase sequence component was absent. This would be true for a three-phase, three-wire system having no neutral conductor, when there could be no residue, and the networks shown may be used without modification. If, however, there exists a neutral conductor, in which case the currents will not necessarily be without residue, and zero-phase sequence components of current will be present, it is necessary to devise means whereby this component can be eliminated.

Two methods of doing this are shown in the circuits of Fig. 41. The first involves the use of two current transformers whose secondary windings are cross-connected. The resultant current of a pair of cross-connected transformers, as shown, will contain only positive- and negative-phase sequence components of the line currents since it will be the vector difference of the two line currents which, as seen previously, can contain no zero-sequence component.

The second method shows the secondary windings of three current transformers connected in delta. Here the zero-sequence components of current will circulate around the delta connection, and the output current will contain only positive- and negative-phase sequence components.

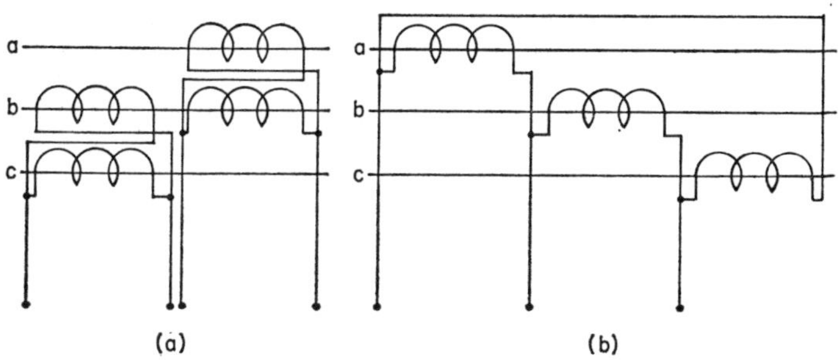

FIG. 41. Circuits for eliminating zero-phase sequence current.

41. Negative-sequence bridge

If unbalanced loading occurs in a three-phase system causing negative-phase sequence currents to be present, serious overheating may occur, as explained previously, in a.c. generators. For this purpose a negative-phase sequence relay may be used in conjunction with a suitable filter network which allows only negative-phase sequence currents to pass through it. This form of protection may also be used on overhead transmission lines against an open-circuited phase or on a.c. motors against single-phase operation or phase reversal.

Such a filter circuit is shown in the form of a bridge network in Fig. 42. Using this network it is necessary that the zero-phase sequence currents should first be eliminated and this may be done by delta connection of current transformers as shown; thus the currents flowing into the bridge will contain only positive- and negative-phase sequence components.

Impedances of R and $R(\frac{1}{2}+j\frac{1}{2}\sqrt{3}) = -h^2R$ are arranged in the arms of the bridge as shown, and the relay of impedance Z_R is connected between the points A and B. Let the current passing through the relay be I_r, and the current in the arm CB be I. Since zero-phase sequence components of current have been eliminated and there is no residue, we have

$$I_b = -(I_a + I_c).$$

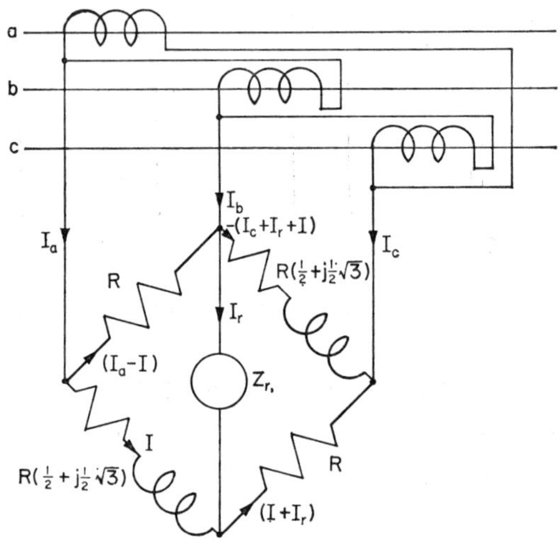

FIG. 42. Negative-phase sequence bridge.

Equating the voltage drops in the two circuits CAD and CBD gives

$$(I_a - I)R + (I_c + I_r + I)h^2R = -h^2IR + (I + I_r)R.$$

Therefore $\quad 2I(h^2 - 1) = -I_a - h^2I_c + I_r(1 - h^2). \qquad (137)$

Equating the voltage drops round the loop $CABC$ gives

$$(I_a - I)R + I_rZ_r + Ih^2R = 0.$$

Therefore $\quad I(h^2 - 1) = -I_a - I_r\dfrac{Z_r}{R}. \qquad (138)$

Eliminating I from eqns. (137) and (138) we have

$$-2I_a - 2I_r \frac{Z_r}{R} = -I_a - h^2 I_c + I_r(1 - h^2).$$

Therefore $$I_r \left(h^2 - 1 - 2\frac{Z_r}{R}\right) = I_a - h^2 I_c.$$

Therefore $$I_r = \frac{I_a - h^2 I_c}{[h^2 - 1 - 2\,(Z_r/R)]},$$

and, since there are no zero-phase sequence components of current,

$$I_r = \frac{(I_{a1} + I_{a2}) - h^2(hI_{a1} + h^2 I_{a2})}{[h^2 - 1 - 2\,(Z_r/R)]},$$

$$= I_{a2} \frac{(1 - h)}{[h^2 - 1 - 2\,(Z_r/R)]}. \qquad (139)$$

As Z_r and R are network constants, the relay current will thus be proportional to the negative-phase sequence current.

42. Positive- and negative-sequence filter

An interesting filter circuit is illustrated in the following example.

EXAMPLE 10. Three equal impedance arms AB, BC, CA are connected in delta to the terminals A, B, C of a 3-phase supply. Each arm consists of two resistors and a capacitor connected in series in the order shown in Fig. 43. The relation between the ohmic values of the components is given by $R{:}r{:}1/\omega C = 2{:}1{:}\sqrt{3}$, and high-impedance voltmeters V_1, V_2 and V_3 are connected to the network as shown. If the supply-line voltages V_{AB}, V_{BC}, C_{CA} are balanced, determine either graphically or algebraically the readings of the three voltmeters (a) when the phase sequence of the supply is A–B–C, and (b) when the phase sequence of the supply is reversed.

Hence explain how the network could be employed to measure, respectively, the positive phase-sequence and the negative phase-sequence voltage components of an asymmetrical 3-phase supply.

[I.E.E., November 1958]

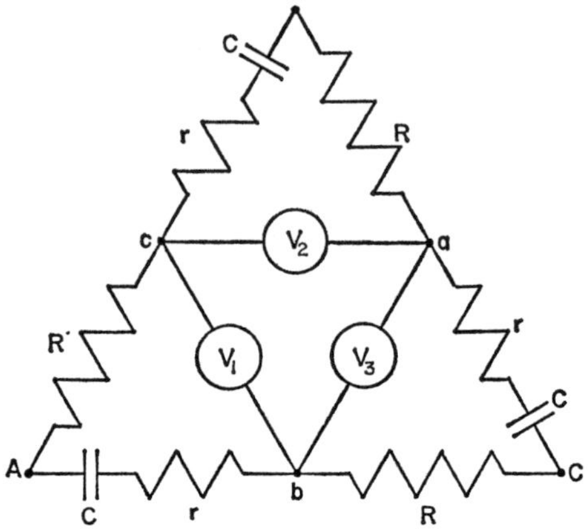

FIG. 43. Example 10.

(1) *Graphical method*

(a) PHASE SEQUENCE *A–B–C*

For the phase sequence *A–B–C* the phase sequence of the line voltages will be V_{AB}, V_{BC}, V_{CA}, as shown in Fig. 44a. Now the voltage across voltmeter V_1 is given by

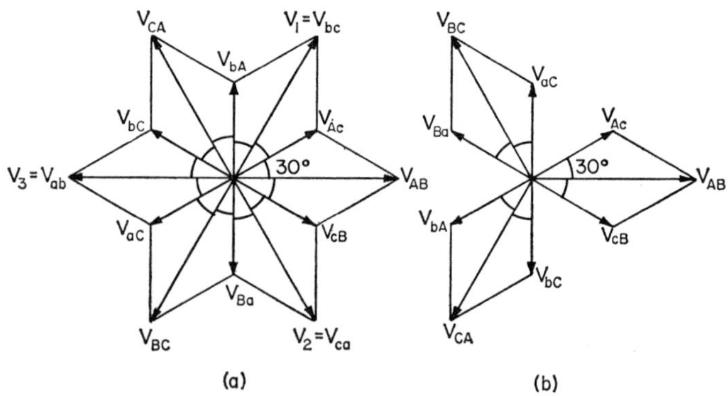

FIG. 44. Example 10—Vector diagrams.

$$V_1 = V_{bc} = V_{bA} + V_{Ac},$$

$$= V_{CA} \frac{\left(r - \dfrac{j}{\omega C}\right)}{\left(R + r - \dfrac{j}{\omega C}\right)} + V_{AB} \frac{R}{\left(R + r - \dfrac{j}{\omega C}\right)},$$

$$= V_{CA} \frac{(1 - j\sqrt{3})}{(3 - j\sqrt{3})} + V_{AB} \frac{2}{(3 - j\sqrt{3})},$$

$$= V_{CA} \left(\tfrac{1}{2} - j\frac{\sqrt{3}}{6}\right) + V_{AB} \left(\tfrac{1}{2} + j\frac{\sqrt{3}}{6}\right), \qquad (140)$$

$$= \frac{1}{\sqrt{3}} V_{CA} \left(\frac{\sqrt{3}}{2} - j\tfrac{1}{2}\right) + \frac{1}{\sqrt{3}} V_{AB} \left(\frac{\sqrt{3}}{2} + j\tfrac{1}{2}\right),$$

$$= \frac{1}{\sqrt{3}} V_{CA} \underline{/-30^\circ} + \frac{1}{\sqrt{3}} V_{AB} \underline{/30^\circ}.$$

Similarly, it may be shown that

$$V_2 = V_{ca} = \frac{1}{\sqrt{3}} V_{AB} \underline{/-30^\circ} + \frac{1}{\sqrt{3}} V_{BC} \underline{/30^\circ},$$

$$V_3 = V_{ab} = \frac{1}{\sqrt{3}} V_{BC} \underline{/-30^\circ} + \frac{1}{\sqrt{3}} V_{CA} \underline{/30^\circ}.$$

From the complete vector diagram it is seen that, considering V_1,

$$V_{Ac} = -V_{aC} \quad \text{and} \quad V_{bA} = -V_{Ba},$$

and so it follows that

$$V_1 = -V_{BC}.$$

Similarly,

$$V_2 = -V_{CA},$$
$$V_3 = -V_{AB}.$$

The three voltmeters will thus each read the magnitude of the line voltage when the phase sequence is A–B–C.

(b) PHASE SEQUENCE REVERSED

With the phase sequence reversed the line voltages may be shown vectorially as in Fig. 44b. Here it is seen that

$$V_{bA} = -V_{Ac}$$

and hence

$$V_1 = V_{bc} = V_{bA} + V_{Ac} = 0.$$

Similarly,

$$V_2 = 0, \quad V_3 = 0,$$

and so each of the voltmeter readings will be zero.

(2) *Algebraic method*

(a) PHASE SEQUENCE *A–B–C*

From eqn. (140) is obtained

$$V_1 = V_{bc} = V_{CA}\left(\tfrac{1}{2} - j\frac{\sqrt{3}}{6}\right) + V_{AB}\left(\tfrac{1}{2} + j\frac{\sqrt{3}}{6}\right).$$

Since the phase sequence is *A–B–C*,

$$V_{CA} = hV_{AB},$$

$$\begin{aligned} V_1 &= hV_{AB}\left(\tfrac{1}{2} - j\frac{\sqrt{3}}{6}\right) + V_{AB}\left(\tfrac{1}{2} + j\frac{\sqrt{3}}{6}\right), \\ &= V_{AB}\left[\left(-\tfrac{1}{2} + j\tfrac{1}{2}\sqrt{3}\right)\left(\tfrac{1}{2} - j\frac{\sqrt{3}}{6}\right) + \left(\tfrac{1}{2} + j\frac{\sqrt{3}}{6}\right)\right], \\ &= V_{AB}\left(\tfrac{1}{2} + j\tfrac{1}{2}\sqrt{3}\right), \\ &= -h^2V_{AB}, \\ &= -V_{BC}, \end{aligned}$$

and similarly

$$V_2 = -V_{CA},$$

$$V_3 = -V_{AB},$$

(b) PHASE SEQUENCE REVERSED

Again, from eqn. (140),

$$V_1 = V_{bc} = V_{CA}\left(\tfrac{1}{2} - j\frac{\sqrt{3}}{6}\right) + V_{AB}\left(\tfrac{1}{2} + j\frac{\sqrt{3}}{6}\right).$$

The phase sequence will now be A–C–B and so

$$V_{CA} = h^2 V_{AB}.$$

Therefore

$$V_1 = h^2 V_{AB}\left(\tfrac{1}{2} - j\frac{\sqrt{3}}{6}\right) + V_{AB}\left(\tfrac{1}{2} + j\frac{\sqrt{3}}{6}\right),$$

$$= V_{AB}\left[\left(-\tfrac{1}{2} - j\tfrac{1}{2}\sqrt{3}\right)\left(\tfrac{1}{2} - j\frac{\sqrt{3}}{6}\right) + \left(\tfrac{1}{2} + j\frac{\sqrt{3}}{6}\right)\right],$$

$$= 0.$$

Similarly,

$$V_2 = 0.$$

$$V_3 = 0.$$

In order to measure the positive-phase sequence voltages of a three-phase supply whose terminals are A^1, B^1, and C^1 and phase sequence A^1–B^1–C^1, the three voltmeters would read the positive-phase sequence component of voltage if A^1 is connected to A, B^1 to B, and C^1 to C of the network. To measure the negative-phase sequence component two of the connections would have to be reversed thus connecting A^1 to A, B^1 to C, and C^1 to B.

43. Zero-sequence power

Using the appropriate filter networks to produce currents and voltages of the correct phase sequence and correct phase displacement relative to one another, the symmetrical components of power may be measured using a dynamometer type wattmeter by passing the currents through the wattmeter coils. The zero-sequence power may be measured by passing zero-phase sequence current through the current coil of the instrument and applying zero-phase sequence voltage across the voltage coil. This may be done by using a combination of the circuits of Figs. 34 and 36 used for measuring the zero-phase sequence voltage and current as

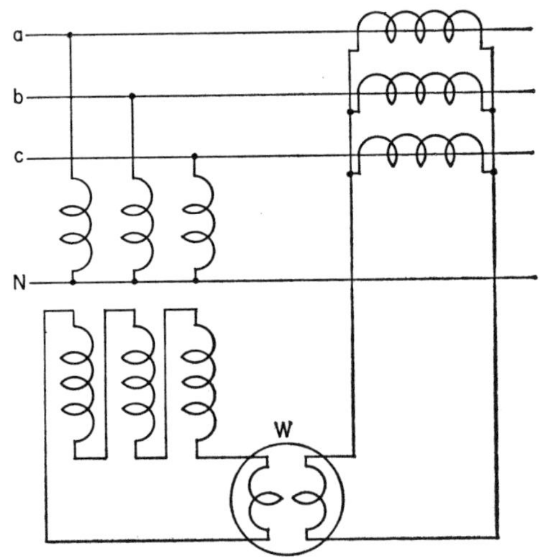

FIG. 45. Measurement of zero-phase sequence power.

shown in Fig. 45. The wattmeter may then be calibrated to measure the total power due to the zero-phase sequence components of voltage and current in the three phases. Since the zero-phase sequence components of both voltage and current are the same in each of the three phases, the wattmeter will have applied to it the correct phase displacement between voltage and current and so no phase correction will be necessary.

44. Positive-sequence power

The positive-phase sequence components of voltage and of current in the three phases have a phase displacement of 120° or

240° between the corresponding quantities for one phase and the next. Because of this it is essential that in measuring the positive-phase sequence power the components of current and voltage fed to the wattmeter must be of the same phase otherwise the phase displacement between current and voltage will be incorrect.

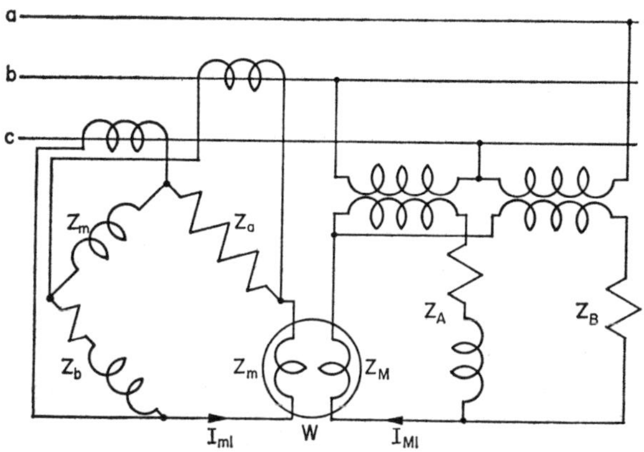

FIG. 46. Measurement of positive-phase sequence power.

With certain adjustments, the circuits shown in Figs. 37 and 40 for measuring positive-phase sequence voltage and current respectively may be combined as shown in Fig. 46 for the operation of a positive-sequence wattmeter.

From eqn. (135) the current I_{m1} flowing through the current coil m of the wattmeter is given by

$$I_{m1} = I_{a1}.$$

The current I_{M1} in the voltage coil would be given from eqn. (126) as

$$I_{M1} = \frac{3V_{a1}}{Z_b + Z_M(1-h)},$$

$$= \frac{3V_{a1}}{R + \sqrt{(3)}\, Z_M \underline{/-30^\circ}},$$

where $Z_b = R$ and is purely resistive.

In order that the phase angle of I_{M1} shall be the same as that of the positive phase sequence voltage V_{a1}, two things are possible. First, the impedance Z_M of the voltage coil may be small as compared with R in which case

$$I_{M1} \simeq \frac{3V_{a1}}{R}.$$

The power measured by the wattmeter may then represented by

$$P + jQ = \frac{3V_{a1}^{*} I_{a1}}{R}. \qquad (141)$$

The wattmeter may then be calibrated to give the correct value of the total positive-sequence power for the three phases.

If, on the other hand, Z_M is not negligible, as compared with the value of R, the power factor angle of Z_M would have to be made 30° lagging when we may write

$$Z_M = R_M \underline{/30^\circ}.$$

Thus we have

$$I_{M1} = \frac{3V_{a1}}{R + \sqrt{(3)}\, R_M},$$

and so the power measured by the wattmeter may be written as

$$P + jQ = \frac{3V_{a1}^{*} I_{a1}}{R + \sqrt{(3)}\, R_M},$$

so that, as in the first case, the wattmeter may again be calibrated to give the positive-sequence power.

45. Negative-sequence power

Just as it is possible to measure positive-sequence power with a single wattmeter, so also can negative-sequence power be measured by supplying negative-phase sequence current to the current coil and applying negative-phase sequence voltage across the voltage coil. This may be done using any of the filter networks already given but care must be taken to maintain the correct phase relationship between the voltage and current applied to the instrument.

Exercises

1. An inductor and a capacitor, each of reactance X, and two resistors each of resistance R are connected to the terminals A, B, C of a balanced 3-phase supply as shown in Fig. 47. A high-impedance voltmeter D is connected between the points a and c. The line voltages are $V_{AB} = V$, $V_{BC} = h^2 V$ and $V_{CA} = hV$ where $h = \frac{1}{2}(-1 + j\sqrt{3})$. Find the ratio X/R for which the voltmeter D reads zero, and determine for this ratio the voltage applied to the voltmeter if the phase sequence of the supply is reversed.

[I.E.E., November 1961]

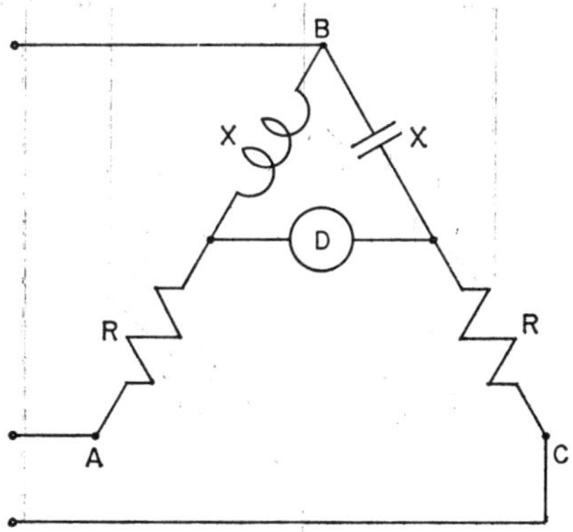

FIG. 47. Exercise 1.

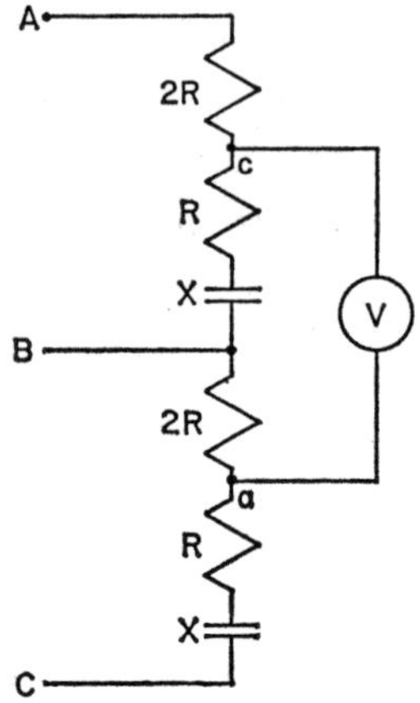

FIG. 48. Exercise 2.

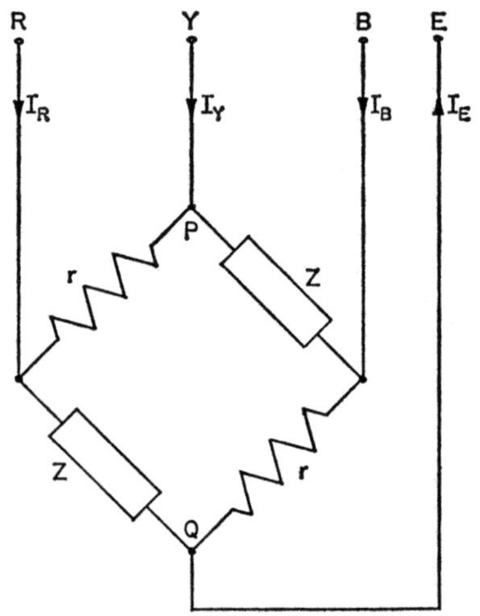

FIG. 49. Exercise 3.

2. Two equal impedance arms AB and BC are connected to the terminals A, B, C of a 3-phase supply, as shown in Fig. 48. Each capacitor has a reactance of $X = \sqrt{(3)}R$. A high impedance voltmeter V is connected to the circuit at the points a and c as shown. If the supply line voltages V_{AB}, V_{BC}, V_{CA}, are balanced, determine either algebraically or graphically the reading of the voltmeter (a) when the phase sequence of the supply voltages is A–B–C, and (b) when the phase sequence is reversed.

Hence explain how the network could be employed to measure, respectively, the positive and the negative-phase-sequence voltage components of an unbalanced 3-phase supply. [I.E.E., June 1961]

3. The bridge circuit shown in Fig. 49 is made up with two equal resistors r and two equal impedances Z as shown, such that $r = 2$ ohms and $Z = (1 + j\sqrt{3})$ ohms. The bridge is supplied with unbalanced currents I_R, I_Y, I_B, from a 3-phase 4-wire symmetrical supply and the phase sequence is RYB. In the R line the positive-, negative- and zero-sequence components of current are, respectively $(0 + j50)$, $5(3 + j\sqrt{3})$ and $(5 + j0)$ amperes. Calculate the potential difference between the points P and Q of the bridge.

[I.E.E., May 1962]

CHAPTER 5

Short-circuit Faults on an Alternating-current Generator

THE use of the theory of symmetrical components is directly applicable to problems involving the simple case of unbalanced short circuits on three-phase a.c. generators. The four cases to be considered are:

(1) Line-to-line fault.
(2) Single line-to-earth fault.
(3) Double line-to-earth fault.
(4) Three-phase fault.

Although the three-phase fault is not an unbalanced condition it is included for the sake of completeness.

Throughout this chapter the symbols E and V with appropriate subscripts will be used to denote generated e.m.f. and terminal voltage respectively. The condition will be considered for a direct fault and also a fault containing an impedance.

In the analysis of each fault condition the following may be considered and made use of:

(1) The terminal conditions for voltage and current.
(2) The equations giving the symmetrical components of current.
(3) The equations giving the symmetrical components of voltage.
(4) Application of Kirchhoff's second law to phase sequence voltage drops.

Whilst analysing each of the fault conditions it should be remembered that the only e.m.f. generated by a three machine is of positive phase sequence, E_{a2} and E_{a0}, both being zero.

46. Line-to-line fault

Figure 50 represents a three-phase star-connected generator having a short circuit across terminals b and c.

Let Z_1, Z_2, and Z_0 be the impedances of the generator to positive-, negative-, and zero-phase sequence currents respectively.

The terminal conditions at the fault are:

$$\left.\begin{aligned} I_a &= 0, \\ I_b &= -I_c, \\ V_b &= V_c. \end{aligned}\right\}$$

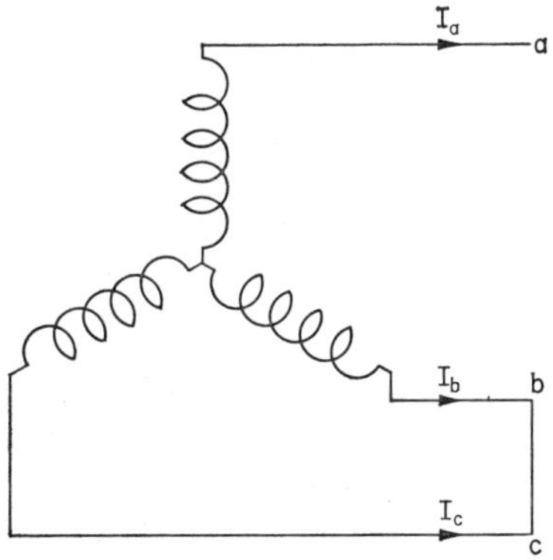

FIG. 50. Line-to-line fault.

The symmetrical components of current are given by:

$$\left.\begin{aligned} I_{a0} &= \tfrac{1}{3}(I_a + I_b + I_c) = 0, \\ I_{a1} &= \tfrac{1}{3}(I_a + hI_b + h^2I_c) = \tfrac{1}{3}(h - h^2)I_b = j\frac{I_b}{\sqrt{3}}, \\ I_{a2} &= \tfrac{1}{3}(I_a + h^2I_b + hI_c) = \tfrac{1}{3}(h^2 - h)I_b = -j\frac{I_b}{\sqrt{3}}. \end{aligned}\right\} \quad (142)$$

Therefore $$I_{a1} = -I_{a2}. \tag{143}$$

The symmetrical components of terminal voltage are given by:

$$\left.\begin{aligned} V_{a0} &= \tfrac{1}{3}(V_a + V_b + V_c),\\ V_{a1} &= \tfrac{1}{3}(V_a + hV_b + h^2V_c) = \tfrac{1}{3}[V_a + (h + h^2)V_b],\\ V_{a2} &= \tfrac{1}{3}(V_a + h^2V_b + hV_c) = \tfrac{1}{3}[V_a + (h + h^2)V_b], \end{aligned}\right\} \tag{144}$$

Therefore $$V_{a1} = V_{a2}. \tag{145}$$

Applying Kirchhoff's second law to the phase sequence voltage drops gives

$$\left.\begin{aligned} V_{a0} &= -I_{a0}Z_0 = 0,\\ V_{a1} &= E_{a1} - I_{a1}Z_1,\\ V_{a2} &= -I_{a2}Z_2. \end{aligned}\right\} \tag{146}$$

Now since $$V_{a1} = V_{a2},$$

we have $$E_{a1} - I_{a1}Z_1 = -I_{a2}Z_2.$$

Therefore $$E_{a1} - j\frac{I_b}{\sqrt{3}}Z_1 = j\frac{I_b}{\sqrt{3}}Z_2.$$

Therefore $$I_b = \frac{-j\sqrt{(3)}E_{a1}}{Z_1 + Z_2}. \tag{147}$$

Therefore $$I_c = \frac{j\sqrt{(3)}E_{a1}}{Z_1 + Z_2}. \tag{148}$$

Also, since $$E_{a1} - I_{a1}Z_1 = -I_{a2}Z_2,$$

$$E_{a1} - I_{a1}Z_1 = I_{a1}Z_2.$$

Therefore $$I_{a1} = \frac{E_{a1}}{Z_1 + Z_2}. \tag{149}$$

Therefore
$$I_{a2} = \frac{-E_{a1}}{Z_1 + Z_2}. \tag{150}$$

And
$$V_{a1} = V_{a2} = -I_{a2} Z_2$$
$$= \frac{Z_2}{Z_1 + Z_2} E_{a1}. \tag{151}$$

The phase voltages are given by:

$$V_a = V_{a0} + V_{a1} + V_{a2},$$
$$= \frac{2Z_2}{Z_1 + Z_2} E_{a1}; \tag{152}$$
$$V_b = V_{a0} + h^2 V_{a1} + h V_{a2},$$
$$= \frac{(h^2 + h)Z_2}{Z_1 + Z_2} E_{a1},$$
$$= \frac{-Z_2}{Z_1 + Z_2} E_{a1}; \tag{153}$$
$$V_c = V_b = \frac{-Z_2}{Z_1 + Z_2} E_{a1}. \tag{154}$$

47. Line-to-line fault through an impedance

Perhaps the most common type of fault is that occurring through an impedance. This is represented in Fig. 51 where the impedance exists between lines. Let Z be the value of the fault impedance, then the terminal conditions at the fault are given by:

$$\left.\begin{aligned} I_a &= 0, \\ I_b &= -I_c, \\ V_b &= V_c - I_c Z. \end{aligned}\right\}$$

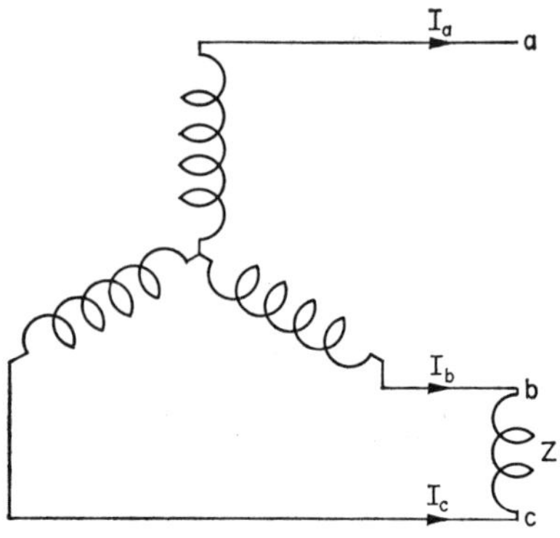

FIG. 51. Line-to-line fault through an impedance.

As in the previous case, the symmetrical components of current will be given by:

$$\left.\begin{aligned} I_{a0} &= 0, \\ I_{a1} &= -I_{a2} = j\frac{I_b}{\sqrt{3}}. \end{aligned}\right\}$$

The symmetrical components of voltage are now given by:

$$\left.\begin{aligned} V_{a0} &= \tfrac{1}{3}(V_a + V_b + V_c), \\ V_{a1} &= \tfrac{1}{3}(V_a + hV_b + h^2V_c) = \tfrac{1}{3}[V_a + (h + h^2)V_c - hI_cZ], \\ V_{a2} &= \tfrac{1}{3}(V_a + h^2V_b + hV_c) = \tfrac{1}{3}[V_a + (h + h^2)V_c - h^2I_cZ]. \end{aligned}\right\} \quad (155)$$

Equations (146) apply as in the previous case.

From eqns. (155) is obtained

$$\begin{aligned} V_{a1} - V_{a2} &= 1/3(h^2 - h)I_cZ, \\ &= 1/3(h^2 - h)(hI_{a1} - h^2I_{a1})Z, \\ &= I_{a1}Z. \end{aligned}$$

Therefore $$V_{a1} = V_{a2} + I_{a1}Z$$

and, substituting from eqn. (146) for V_{a2}, gives

$$V_{a1} = -I_{a2}Z_2 + I_{a1}Z,$$
$$= I_{a1}(Z_2 + Z) \quad (\text{since } I_{a1} = -I_{a2}).$$

Substituting for V_{a1} from eqns. (146) we have

$$E_{a1} - I_{a1}Z_1 = I_{a1}(Z_2 + Z).$$

Therefore $$I_{a1} = \frac{E_{a1}}{Z_1 + Z_2 + Z} \tag{156}$$

and $$I_{a2} = \frac{-E_{a1}}{Z_1 + Z_2 + Z}. \tag{157}$$

Having determined the symmetrical components of current the phase currents I_b and I_c are given by

$$I_b = I_{a0} + h^2 I_{a1} + hI_{a2},$$
$$= \frac{(h^2 - h)E_{a1}}{Z_1 + Z_2 + Z},$$
$$= \frac{-j\sqrt{(3)}\,E_{a1}}{Z_1 + Z_2 + Z}. \tag{158}$$

Therefore $$I_c = \frac{j\sqrt{(3)}\,E_{a1}}{Z_1 + Z_2 + Z}. \tag{159}$$

The symmetrical components of voltage are now

$$V_{a1} = I_{a1}(Z_2 + Z),$$
$$= \frac{(Z_2 + Z)}{Z_1 + Z_2 + Z}E_{a1}; \tag{160}$$
$$V_{a2} = -I_{a2}Z_2,$$
$$= \frac{Z_2E_{a1}}{Z_1 + Z_2 + Z}, \tag{161}$$

and from eqn. (159)

$$V_{a0} = 0. \tag{162}$$

Hence the phase voltages are given by

$$V_a = V_{a0} + V_{a1} + V_{a2},$$
$$= \frac{(2Z_2 + Z)E_{a1}}{Z_1 + Z_2 + Z}; \qquad (163)$$
$$V_b = V_{a0} + h^2V_{a1} + hV_{a2},$$
$$= \frac{(h^2Z - Z_2)E_{a1}}{Z_1 + Z_2 + Z}. \qquad (164)$$
$$V_c = V_{a0} + hV_{a1} + h^2V_{a2},$$
$$= \frac{(hZ - Z_2)E_{a1}}{Z_1 + Z_2 + Z}. \qquad (165)$$

48. Single line-to-earth fault

A generator with earthed neutral having a single line-to-earth fault on phase a is represented in Fig. 52. The terminal conditions are given by

$$\left.\begin{aligned} I_b &= 0, \\ I_c &= 0, \\ V_a &= 0. \end{aligned}\right\}$$

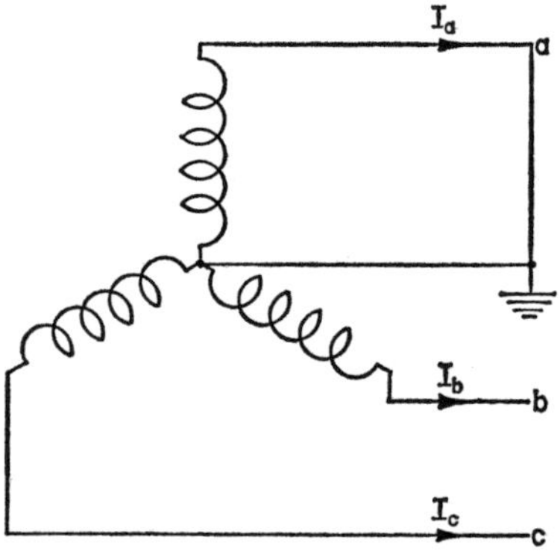

FIG. 52. Single line-to-earth fault.

Applying these terminal conditions to the expressions for the symmetrical components of current gives

$$\left.\begin{aligned} I_{a0} &= \tfrac{1}{3}(I_a + I_b + I_c) = \tfrac{1}{3}I_a, \\ I_{a1} &= \tfrac{1}{3}(I_a + hI_b + h^2 I_c) = \tfrac{1}{3}I_a, \\ I_{a2} &= \tfrac{1}{3}(I_a + h^2 I_b + hI_c) = \tfrac{1}{3}I_a. \end{aligned}\right\}$$

i.e. $$I_{a0} = I_{a1} = I_{a2} = \tfrac{1}{3}I_a. \tag{166}$$

From eqns. (146) we have

$$\left.\begin{aligned} V_{a0} &= -\,I_{a0}Z_0 = -\,\tfrac{1}{3}I_a Z_0, \\ V_{a1} &= E_{a1} - I_{a1}Z_1 = E_{a1} - \tfrac{1}{3}I_a Z_1, \\ V_{a2} &= -\,I_{a2}Z_2 = -\,\tfrac{1}{3}I_a Z_2. \end{aligned}\right\} \tag{167}$$

And since

$$V_a = 0$$

then

$$V_{a0} + V_{a1} + V_{a2} = 0.$$

Therefore $$-\,\tfrac{1}{3}I_a Z_0 + E_{a1} - \tfrac{1}{3}I_a Z_1 - \tfrac{1}{3}I_a Z_2 = 0,$$

$$I_a = \frac{3E_{a1}}{Z_0 + Z_1 + Z_2}. \tag{168}$$

Hence $$I_{a0} = I_{a1} = I_{a2} = \tfrac{1}{3}I_a = \frac{E_{a1}}{Z_0 + Z_1 + Z_2}. \tag{169}$$

The symmetrical components of voltage are given by

$$V_{a0} = \frac{-\,E_{a1}Z_0}{Z_0 + Z_1 + Z_2}, \tag{170}$$

$$V_{a1} = \frac{E_{a1}\,(Z_0 + Z_2)}{Z_0 + Z_1 + Z_2}, \tag{171}$$

$$V_{a2} = \frac{-\,E_{a1}Z_2}{Z_0 + Z_1 + Z_2}. \tag{172}$$

By definition $V_a = 0$ and the phase voltages V_b and V_c are given by

$$V_b = V_{a0} + h^2 V_{a1} + hV_{a2},$$

$$= \frac{-E_{a1}Z_0}{Z_0+Z_1+Z_2} + \frac{h^2 E_{a1}(Z_0+Z_2)}{Z_0+Z_1+Z_2} - \frac{hE_{a1}Z_2}{Z_0+Z_1+Z_2},$$

$$= \frac{E_{a1}\left[(h^2-1)Z_0 + (h^2-h)Z_2\right]}{Z_0+Z_1+Z_2},$$

$$= \frac{-j\sqrt{(3)}\,E_{a1}(Z_2 - hZ_0)}{Z_0+Z_1+Z_2}; \qquad (173)$$

$$V_c = V_{a0} + hV_{a1} + h^2 V_{a2},$$

$$= \frac{-E_{a1}Z_0}{Z_0+Z_1+Z_2} + \frac{hE_{a1}(Z_0+Z_2)}{Z_0+Z_1+Z_2} - \frac{h^2E_{a1}Z_2}{Z_0+Z_1+Z_2},$$

$$= \frac{E_{a1}\left[(h-1)Z_0 + (h-h^2)Z_2\right]}{Z_0+Z_1+Z_2},$$

$$= \frac{j\sqrt{(3)}\,E_{a1}\,(Z_2 - h^2Z_0)}{Z_0+Z_1+Z_2}. \qquad (174)$$

49. Single line-to-earth fault through an impedance

This condition is illustrated in Fig. 53, where the fault impedance Z is connected between phase a and earth. With the neutral of the generator solidly earthed the following terminal conditions apply:

$$\left.\begin{aligned} I_b &= 0, \\ I_c &= 0, \\ V_a &= I_a Z. \end{aligned}\right\}$$

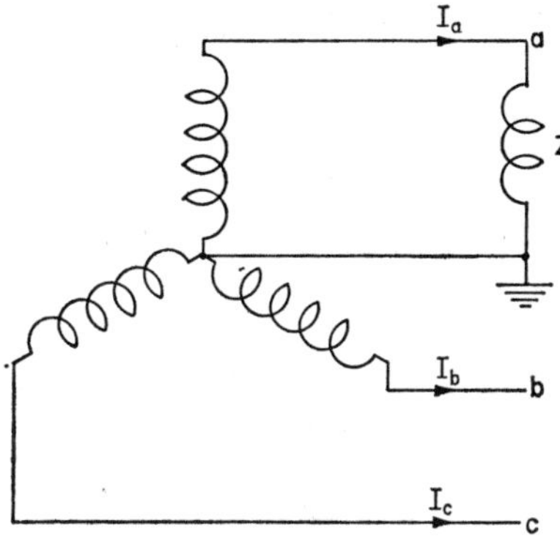

FIG. 53. Single line-to-earth fault through an impedance.

As in the previous case of a line-to-earth fault without an impedance in the fault the symmetrical components of current are again given by eqns. (166) where

$$I_{a0} = I_{a1} = I_{a2} = \tfrac{1}{3}I_a$$

and from eqns. (167) the symmetrical components of voltage at the fault are given by

$$\left.\begin{aligned} V_{a0} &= -\tfrac{1}{3}I_aZ_0, \\ V_{a1} &= E_{a1} - \tfrac{1}{3}I_aZ_1, \\ V_{a2} &= -\tfrac{1}{3}I_aZ_2. \end{aligned}\right\}$$

Now from the terminal conditions we have

$$\begin{aligned} V_a = I_aZ &= V_{a0} + V_{a1} + V_{a2} \\ &= -\tfrac{1}{3}I_aZ_0 + E_{a1} - \tfrac{1}{3}I_aZ_1 - \tfrac{1}{3}I_aZ_2. \end{aligned}$$

Therefore $$I_a = \frac{3E_{a1}}{Z_0 + Z_1 + Z_2 + 3Z}. \tag{175}$$

Thus the symmetrical components of current are given by

$$I_{a0} = I_{a1} = I_{a2} = \frac{E_{a1}}{Z_0 + Z_1 + Z_2 + 3Z}. \tag{176}$$

Using eqns. (167) the symmetrical components of voltage are then given by

$$V_{a0} = \frac{-E_{a1}Z_0}{Z_0 + Z_1 + Z_2 + 3Z}, \tag{177}$$

$$V_{a1} = \frac{E_{a1}(Z_0 + Z_2 + 3Z)}{Z_0 + Z_1 + Z_2 + 3Z}, \tag{178}$$

$$V_{a2} = \frac{-E_{a1}Z_2}{Z_0 + Z_1 + Z_2 + 3Z}. \tag{179}$$

The three terminal voltages may now be found where from the terminal conditions

$$\begin{aligned} V_a &= I_a Z, \\ &= \frac{3E_{a1}Z}{Z_0 + Z_1 + Z_2 + 3Z}. \end{aligned} \tag{180}$$

And from the symmetrical components we have

$$\begin{aligned} V_b &= V_{a0} + h^2 V_{a1} + hV_{a2}, \\ &= \frac{-E_{a1}Z_0 + h^2 E_{a1}(Z_0 + Z_2 + 3Z) - hE_{a1}Z_2}{Z_0 + Z_1 + Z_2 + 3Z}, \\ &= \frac{E_{a1}[(h^2 - 1)Z_0 + (h^2 - h)Z_2 + 3h^2 Z]}{Z_0 + Z_1 + Z_2 + 3Z}, \\ &= \frac{E_{a1}[3h^2 Z - j\sqrt{(3)}(Z_2 - hZ_0)]}{Z_0 + Z_1 + Z_2 + 3Z}, \end{aligned} \tag{181}$$

and $\quad V_c = V_{a0} + hV_{a1} + h^2V_{a2},$

$$= \frac{-E_{a1}Z_0 + hE_{a1}(Z_0 + Z_2 + 3Z) - h^2E_{a1}Z_2}{Z_0 + Z_1 + Z_2 + 3Z},$$

$$= \frac{E_{a1}[(h-1)Z_0 + (h-h^2)Z_2 + 3hZ]}{Z_0 + Z_1 + Z_2 + 3Z},$$

$$= \frac{E_{a1}[3hZ + j\sqrt{(3)}(Z_2 - h^2Z_0)]}{Z_0 + Z_1 + Z_2 + 3Z}. \tag{182}$$

EXAMPLE 11. A 6·6 kV generator with star-connected windings has its star point earthed. Its positive-, negative-, and zero-sequence impedances are respectively 1·5, 1·0, and 0·5 Ω. The generator feeds a three-phase cable on which a fault occurs at the far end of the cable between one conductor and the sheath.

If the positive-, negative-, and zero-sequence impedances of the cable are 0·8, 0·8, and 2·5 Ω respectively, find the fault current and the phase voltage of the generator on the faulty phase.

Using the subscripts G and L to signify the phase-sequence impedances of the generator and line respectively we have

$$\left.\begin{aligned} Z_{G0} &= j\,0{\cdot}5, \\ Z_{G1} &= j\,1{\cdot}5, \\ Z_{G2} &= j\,1{\cdot}5, \end{aligned}\right\}$$

and

$$\left.\begin{aligned} Z_{L0} &= j\,2{\cdot}5, \\ Z_{L1} &= j\,0{\cdot}8, \\ Z_{L2} &= j\,0{\cdot}8. \end{aligned}\right\}$$

Now 6·6 kV is the generator line voltage and so the phase voltage will be given by

$$\frac{6{,}600}{\sqrt{3}} = 3810\ V.$$

Using eqn. (168) gives

$$I_a = \frac{3E_{a1}}{Z_0 + Z_1 + Z_2},$$

$$= \frac{3E_{a1}}{(Z_{G0} + Z_{L0}) + (Z_{G1} + Z_{L1}) + (Z_{G2} + Z_{L2})},$$

$$= \frac{3 \times 3810}{(j0\cdot 5 + j2\cdot 5) + (j1\cdot 5 + j0\cdot 8) + (j1\cdot 0 + j0\cdot 8)},$$

$$= \frac{-j\,11{,}430}{7\cdot 1},$$

$$= -j2030 \equiv 2030 \text{ A.}$$

Let the generator terminals be represented by a', b', and c' in Fig. 54, then the terminal voltage V'_a of the faulty phase is given by

$$V'_a = V'_{a0} + V'_{a1} + V'_{a2},$$

$$= I_{a0}Z_{L0} + I_{a1}Z_{L1} + I_{a2}Z_{L2},$$

$$= \frac{I_a}{3}(Z_{L0} + Z_{L1} + Z_{L2}),$$

$$= \frac{-j\,2030}{3}(j2\cdot 5 + j0\cdot 8 + j0\cdot 8) \equiv 2780 \text{ V.}$$

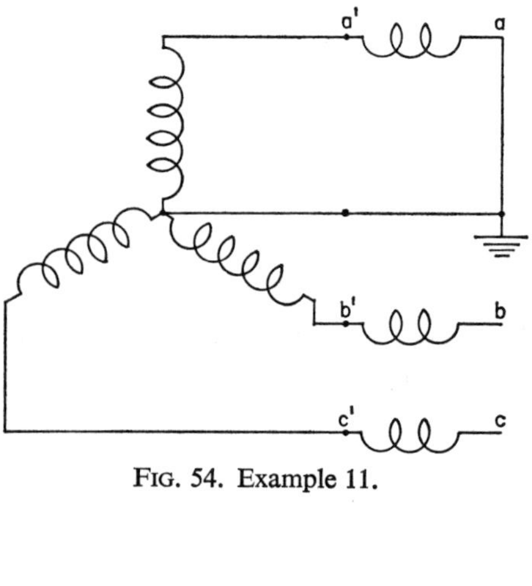

FIG. 54. Example 11.

50. Double line-to-earth fault

A generator having a line-to-line-earth fault is shown in Fig. 55 with an earthed neutral and no impedance in the fault.

The terminal conditions at the fault are

$$\left.\begin{aligned} I_a &= 0, \\ V_b &= 0, \\ V_c &= 0. \end{aligned}\right\}$$

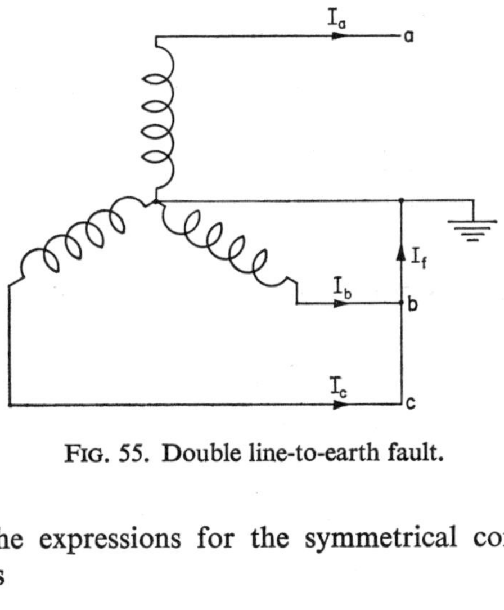

FIG. 55. Double line-to-earth fault.

Writing the expressions for the symmetrical components of voltage gives

$$\left.\begin{aligned} V_{a0} &= \tfrac{1}{3}(V_a + V_b + V_c) = \tfrac{1}{3}V_a, \\ V_{a1} &= \tfrac{1}{3}(V_a + hV_b + h^2V_c) = \tfrac{1}{3}V_a, \\ V_{a2} &= \tfrac{1}{3}(V_a + h^2V_b + hV_c) = \tfrac{1}{3}V_a, \end{aligned}\right\}$$

Therefore $V_{a0} = V_{a1} = V_{a2} = \tfrac{1}{3}V_a.$ (183)

Since the current in phase a is zero we have

$$I_{a0} + I_{a1} + I_{a2} = 0. \tag{184}$$

Expressing V_b in terms of its symmetrical components gives

$$V_b = V_{a0} + h^2 V_{a1} + hV_{a2} = 0$$

from the terminal conditions.

Substituting the expressions for V_{a0}, V_{a1}, and V_{a2} of eqn. (146) leads to

$$-I_{a0}Z_0 + h^2E_{a1} - h^2I_{a1}Z_1 - hI_{a2}Z_2 = 0.$$

Therefore
$$I_{a0}Z_0 + h^2I_{a1}Z_1 + hI_{a2}Z_2 = h_2E_{a1}. \tag{185}$$

Also, expressing V_c in terms of its symmetrical components, gives

$$V_c = V_{a0} + hV_{a1} + h^2V_{a2} = 0,$$

and, again using the expressions of eqn. (146), for V_{a0}, V_{a1}, and V_{a2} gives

$$-I_{a0}Z_0 + hE_{a1} - hI_{a1}Z_1 - h^2I_{a2}Z_2 = 0.$$

Therefore
$$I_{a0}Z_0 + hI_{a1}Z_1 + h^2I_{a2}Z_2 = hE_{a1}. \tag{186}$$

Solving eqns. (184), (185), and (186) simultaneously the following results are obtained for the symmetrical components of current:

$$I_{a0} = \frac{-E_{a1}Z_2}{Z_0Z_1 + Z_1Z_2 + Z_2Z_0},$$

$$= \frac{-E_{a1}}{Z_1 + (Z_0Z_2)/(Z_0 + Z_2)} \cdot \frac{Z_2}{Z_0 + Z_2}; \tag{187}$$

$$I_{a1} = \frac{E_{a1}(Z_0 + Z_2)}{Z_0Z_1 + Z_1Z_2 + Z_2Z_0},$$

$$= \frac{E_{a1}}{Z_1 + (Z_0Z_2)/(Z_0 + Z_2)}; \tag{188}$$

$$I_{a2} = \frac{-E_{a1}Z_0}{Z_0Z_1 + Z_1Z_2 + Z_2Z_0},$$

$$= \frac{-E_{a1}}{Z_1 + (Z_0Z_2)/(Z_0 + Z_2)} \cdot \frac{Z_0}{Z_0 + Z_2}. \tag{189}$$

Having determined the symmetrical components of the current, the phase currents I_b and I_c are given by

$$\begin{aligned} I_b &= I_{a0} + h^2 I_{a1} + hI_{a2}, \\ &= \frac{E_{a1}(-Z_2 + h^2 Z_0 + h^2 Z_2 - hZ_0)}{Z_0 Z_1 + Z_1 Z_2 + Z_2 Z_0}, \\ &= \frac{E_{a1}[(h^2 - h)Z_0 + (h^2 - 1)Z_2]}{Z_0 Z_1 + Z_1 Z_2 + Z_2 Z_0}, \\ &= \frac{-j\sqrt{(3)}\, E_{a1}(Z_0 - hZ_2)}{Z_0 Z_1 + Z_1 Z_2 + Z_2 Z_0}, \end{aligned} \tag{190}$$

and

$$\begin{aligned} I_c &= I_{a0} + hI_{a1} + h^2 I_{a2}, \\ &= \frac{E_{a1}(-Z_2 + hZ_0 + hZ_2 - h^2 Z_0)}{Z_0 Z_1 + Z_1 Z_2 + Z_2 Z_0}, \\ &= \frac{E_{a1}[(h - h^2)Z_0 + (h - 1)Z_2]}{Z_0 Z_1 + Z_1 Z_2 + Z_2 Z_0}, \\ &= \frac{j\sqrt{(3)}\, E_{a1}(Z_0 - h^2 Z_2)}{Z_0 Z_1 + Z_1 Z_2 + Z_2 Z_0}. \end{aligned} \tag{191}$$

If the fault current flowing to earth is I_f, then

$$\begin{aligned} I_f &= I_b + I_c, \\ &= \frac{-3E_{a1} Z_2}{Z_0 Z_1 + Z_1 Z_2 + Z_2 Z_0} = 3I_{a0}. \end{aligned} \tag{192}$$

From eqn. (146)

$$\begin{aligned} V_{a0} &= -I_{a0} Z_0, \\ &= \frac{E_{a1} Z_0 Z_2}{Z_0 Z_1 + Z_1 Z_2 + Z_2 Z_0}, \\ &= \frac{E_{a1}}{Z_1 + (Z_0 Z_2)/(Z_0 + Z_2)} \cdot \frac{Z_0 Z_2}{Z_0 + Z_2}, \end{aligned}$$

and since from eqn. (183)

$$V_{a0} = V_{a1} = V_{a2},$$

we have

$$V_{a0} = V_{a1} = V_{a2} = \frac{E_{a1}}{Z_1 + (Z_0Z_2)/(Z_0 + Z_2)} \cdot \frac{Z_0Z_2}{Z_0 + Z_2}. \tag{193}$$

Also from eqn. (183)

$$V_a = 3V_{a0},$$

$$= \frac{3E_{a1}\,Z_0Z_2}{Z_0Z_1 + Z_1Z_2 + Z_2Z_0}. \tag{194}$$

51. Double line-to-earth fault through an impedance

This condition is represented in Fig. 56 where an impedance Z is contained in the fault to earth.

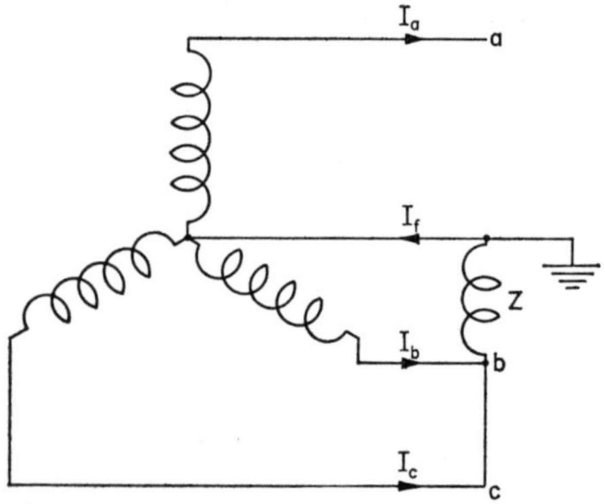

FIG. 56. Double line-to-earth fault through an impedance.

The terminal conditions at the fault are:

$$\left.\begin{aligned} I_a &= 0, \\ V_b &= (I_b + I_c)Z, \\ V_c &= (I_b + I_c)Z. \end{aligned}\right\}$$

The expressions for the symmetrical components of voltage gives

$$\left.\begin{aligned} V_{a0} &= \tfrac{1}{3}(V_a + V_b + V_c) = \tfrac{1}{3}(V_a + 2V_b), \\ V_{a1} &= \tfrac{1}{3}(V_a + hV_b + h^2V_c) = \tfrac{1}{3}(V_a - V_b), \\ V_{a2} &= \tfrac{1}{3}(V_a + h^2V_b + hV_c) = \tfrac{1}{3}(V_a - V_b). \end{aligned}\right\}$$

Therefore $$V_{a1} = V_{a2}. \tag{195}$$

Also since the current I_a is zero it follows that

$$I_{a0} + I_{a1} + I_{a2} = 0. \tag{196}$$

Reverting to the terminal conditions,

$$\begin{aligned} V_b = V_c &= (I_b + I_c)Z, \\ &= [2I_{a0} + (h + h^2)(I_{a1} + I_{a2})]Z, \\ &= [2I_{a0} - (I_{a1} + I_{a2})]Z. \end{aligned} \tag{197}$$

Expressing V_b in terms of its symmetrical components gives

$$V_b = V_{a0} + h^2V_{a1} + hV_{a2} = [2I_{a0} - (I_{a1} + I_{a2})]Z$$

from the terminal conditions.

Substituting for V_{a0}, V_{a1}, and V_{a2} from eqn. (146) we have

$$- I_{a0}Z_0 + h^2E_{a1} - h^2I_{a1}Z_1 - hI_{a2}Z_2 = [2I_{a0} - (I_{a1} + I_{a2})]Z.$$

Therefore

$$I_{a0}(2Z + Z_0) + I_{a1}(h^2Z_1 - Z) + I_{a2}(hZ_2 - Z) = h^2E_{a1}. \tag{198}$$

Also expressing V_c in terms of its symmetrical components gives

$$V_c = V_{a0} + hV_{a1} + h^2V_{a2} = [2I_{a0} - (I_{a1} + I_{a2})]Z$$

and substituting for V_{a0}, V_{a1}, and V_{a2} from eqn. (146) we have

$$-I_{a0}Z_0 + hE_{a1} - hI_{a1}Z_1 - h^2I_{a2}Z_2 = [2I_{a0} - (I_{a1} + I_{a2})]Z.$$

Therefore

$$I_{a0}(2Z + Z_0) + I_{a1}(hZ_1 - Z) + I_{a2}(h^2Z_2 - Z) = hE_{a1}. \tag{199}$$

Solving eqns. (196), (198), and (199) simultaneously gives, for the symmetrical components of current,

$$I_{a0} = \frac{-E_{a1}}{Z_1 + [Z_2(Z_0 + 3Z)]/(Z_0 + Z_2 + 3Z)} \cdot \frac{Z_2}{Z_0 + Z_2 + 3Z}, \tag{200}$$

$$I_{a1} = \frac{E_{a1}}{Z_1 + [Z_2(Z_0 + 3Z)]/(Z_0 + Z_2 + 3Z)}, \tag{201}$$

$$I_{a2} = \frac{-E_{a1}}{Z_1 + [Z_2(Z_0 + 3Z)]/(Z_0 + Z_2 + 3Z)} \cdot \frac{Z_0 + 3Z}{Z_0 + Z_2 + 3Z}. \tag{202}$$

The currents in phases b and c are then found to be

$$\begin{aligned} I_b &= I_{a0} + h^2I_{a1} + hI_{a2}, \\ &= \frac{-j\sqrt{(3)}\,E_{a1}\,(Z_0 - hZ_2 + 3Z)}{Z_1Z_2 + (Z_1 + Z_2)\,(Z_0 + 3Z)}; \end{aligned} \tag{203}$$

$$\begin{aligned} I_c &= I_{a0} + hI_{a1} + h^2I_{a2}, \\ &= \frac{j\sqrt{(3)}\,E_{a1}\,(Z_0 - h^2Z_2 + 3Z)}{Z_1Z_2 + (Z_1 + Z_2)\,(Z_0 + 3Z)}. \end{aligned} \tag{204}$$

The fault current I_f flowing to earth is given by

$$\begin{aligned} I_f &= I_b + I_c, \\ &= \frac{-3\,E_{a1}\,Z_2}{Z_1\,Z_2 + (Z_1 + Z_2)\,(Z_0 + 3Z)} = 3I_{a0}. \end{aligned} \tag{205}$$

The symmetrical components of voltage are

$$V_{a0} = -I_{a0}Z_0,$$

$$= \frac{E_{a1}}{Z_1 + [Z_2(Z_0 + 3Z)]/(Z_0 + Z_2 + 3Z)} \cdot \frac{Z_0 Z_2}{Z_0 + Z_2 + 3Z}; \tag{206}$$

$$V_{a1} = V_{a2} = -I_{a2}Z_2,$$

$$= \frac{E_{a1}}{Z_1 + [Z_2(Z_0 + 3Z)]/(Z_0 + Z_2 + 3Z)} \cdot \frac{Z_2(Z_0 + 3Z)}{Z_0 + Z_2 + 3Z}. \tag{207}$$

The phase voltages are then

$$V_a = V_{a0} + V_{a1} + V_{a2},$$

$$= \frac{3E_{a1}Z_2(Z_0 + 2Z)}{Z_1Z_2 + (Z_1 + Z_2)(Z_0 + 3Z)}, \tag{208}$$

and also

$$V_b = V_c = (I_b + I_c)Z,$$

$$= I_f Z,$$

$$= \frac{-3E_{a1}Z_2Z}{Z_1Z_2 + (Z_1 + Z_2)(Z_0 + 3Z)}. \tag{209}$$

EXAMPLE 12. A 60 MVA 33kV star-connected alternator has positive-, negative- and zero-sequence reactances of 0·3, 0·25 and 0·2 per unit, respectively. The star point of the alternator is earthed through an 8 Ω resistor. A line-line-earth fault occurs on two terminals of the machine; determine the power which will be dissipated in the earthing resistor and also the voltage to earth of the unfaulted terminal of the machine. [I.E.E., 1958]

Since the voltage of 33 kV is a line-to-line voltage the phase voltage will be

$$V_{ph} = \frac{33{,}000}{\sqrt{3}} = 19{,}100 \text{ V}.$$

Also the full load current I is given by

$$I = \frac{60 \times 10^6}{33{,}000 \times \sqrt{3}} = 1050 \text{ A}.$$

The value of the earthing resistor is given in ohms. Converting this to a per unit value gives

$$R_{p.u.} = \frac{1050 \times 8}{19{,}100} = 0 \cdot 44 \text{ p.u.}$$

The calculations may now be performed in terms of per unit quantities. Using eqn. (205) the fault current to earth I_f is given by

$$\begin{aligned} I_f &= \frac{-3E_{a1}Z_2}{Z_1Z_2 + (Z_1 + Z_2)(Z_0 + 3Z)}, \\ &= \frac{-3 \times 1 \times j0 \cdot 25}{(j0 \cdot 3 \times j0 \cdot 25) + (j0 \cdot 3 + j0 \cdot 25)(j0 \cdot 2 + 3 \times 0 \cdot 44)}, \\ &= \frac{j0 \cdot 75}{0 \cdot 185 - j0 \cdot 726}, \\ &= -0 \cdot 97 + j0 \cdot 247. \end{aligned}$$

Therefore

$$\begin{aligned} |I_f| &= \sqrt{(0 \cdot 97^2 + 0 \cdot 247^2)}, \\ &= 1 \cdot 0 \text{ per unit}, \\ &= 1050 \text{ A}. \end{aligned}$$

$$\text{Loss in resistor} = \frac{1050^2 \times 8}{10^6},$$

$$= 8 \cdot 82 \text{ MW}.$$

The voltage to neutral of the unfaulted phase is given in eqn. (208) by

$$V_a = \frac{3E_{a1} Z_2(Z_0 + 2Z)}{Z_1Z_2 + (Z_1 + Z_2)(Z_0 + 3Z)},$$

$$= - I_f (Z_0 + 2Z).$$

The voltage between neutral and earth is

$$V_n = - I_f Z.$$

Hence the voltage to earth of the unfaulted phase is given by V_{ae}, where

$$V_{ae} = - I_f(Z_0 + 3Z) = (0 \cdot 97 - j0 \cdot 247)(j0 \cdot 2 + 3 \times 0 \cdot 44),$$

$$= 1 \cdot 33 - j0 \cdot 132.$$

Therefore $$| V_{ae} | = \sqrt{(1 \cdot 33^2 + 0 \cdot 132^2)},$$

$$= 1 \cdot 335 \text{ p.u.},$$

$$= 19{,}100 \times 1 \cdot 335,$$

$$= 25{,}500 \text{ V}.$$

52. Three-phase fault

In Fig. 57 is shown a generator with a fault across all three phases. The terminal conditions are such that

$$\left.\begin{aligned} V_a = V_b = V_c, \\ I_a + I_b + I_c = 0. \end{aligned}\right\}$$

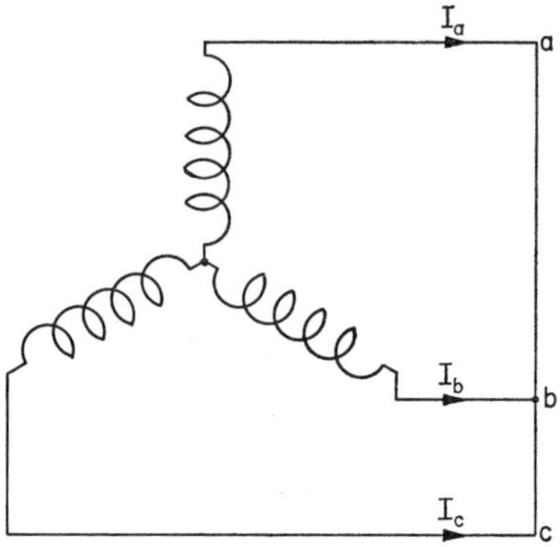

FIG. 57. Three-phase fault.

The expressions for the symmetrical components of voltage give

$$\left.\begin{aligned} V_{a0} &= \tfrac{1}{3}(V_a + V_b + V_c) = V_a, \\ V_{a1} &= \tfrac{1}{3}(V_a + hV_b + h^2V_c) = \tfrac{1}{3}(1 + h + h^2)V_a = 0, \\ V_{a2} &= \tfrac{1}{3}(V_a + h^2V_b + hV_c) = \tfrac{1}{3}(1 + h + h^2)V_a = 0. \end{aligned}\right\} \tag{210}$$

Also

$$I_{a0} = \tfrac{1}{3}(I_a + I_b + I_c) = 0. \tag{211}$$

Now from eqns. (146)

$$V_{a1} = E_{a1} - I_{a1}Z_1.$$

Therefore $\quad 0 = E_{a1} - I_{a1}Z_1$

Therefore $$I_{a1} = \frac{E_{a1}}{Z_1} \tag{212}$$

and $\quad V_{a2} = - I_{a2}Z_2.$

Therefore $$0 = - I_{a2}Z_2. \tag{213}$$

Therefore $\quad I_{a2} = 0.$

Expressing the phase currents in terms of their symmetrical components gives

$$I_a = I_{a0} + I_{a1} + I_{a2},$$
$$= 0 + \frac{E_{a1}}{Z_1} + 0,$$
$$= \frac{E_{a1}}{Z_1}; \tag{214}$$

$$I_b = I_{a0} + h^2 I_{a1} + h I_{a2},$$
$$= 0 + \frac{h^2 E_{a1}}{Z_1} + 0,$$
$$= \frac{h^2 E_{a1}}{Z_1}, \tag{215}$$

and

$$I_c = I_{a0} + h I_{a1} + h^2 I_{a2},$$
$$= 0 + \frac{h E_{a1}}{Z_1} + 0,$$
$$= \frac{h E_{a1}}{Z_1}. \tag{216}$$

The three-phase fault gives rise to no negative- and zero-sequence components in the fault current since the system remains balanced and so could easily be solved by the method normally used in solving balanced three-phase circuits. The same result would then ensue.

53. Three-phase fault through an impedance

It is assumed in this case that the fault impedance Z is the same in all three phases as indicated in Fig. 58. As in the previous case, the system is perfectly balanced and so may be solved in the manner usual for balanced three-phase circuits.

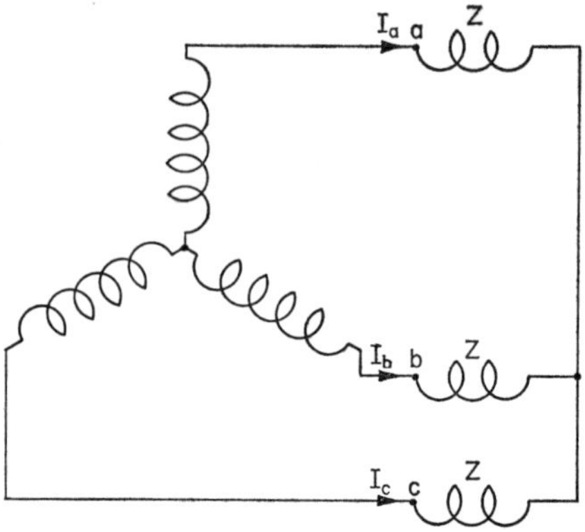

FIG. 58. Three-phase fault through an impedance.

The only current circulating will be of positive phase sequence and so

$$E_{a1} = I_{a1}(Z_1 + Z),$$

$$I_{a1} = \frac{E_{a1}}{Z_1 + Z}. \tag{217}$$

The current flowing in phase a will then be I_{a1}.

Therefore
$$I_a = I_{a1} = \frac{E_{a1}}{Z_1 + Z} \tag{218}$$

and it then follows that

$$I_b = h^2 I_a = \frac{h^2 E_{a1}}{Z_1 + Z} \tag{219}$$

and
$$I_c = h I_a = \frac{h E_{a1}}{Z_1 + Z}. \tag{220}$$

Since there is no unbalance the negative- and zero-phase sequence components of currents are both zero, hence

$$I_{a2} = I_{a0} = 0. \tag{221}$$

The positive-phase sequence voltage at the fault V_{a1} is given by

$$V_{a1} = I_{a1} Z. \tag{222}$$

The negative- and zero-phase sequence components of voltage will both be zero, hence

$$V_{a2} = V_{a0} = 0. \tag{223}$$

The phase voltages at the fault will be given by

$$V_a = V_{a1} = I_{a1}Z,$$

$$= \frac{E_{a1}Z}{Z_1 + Z}; \tag{224}$$

$$V_b = h^2 V_a = \frac{h^2 E_{a1} Z}{Z_1 + Z}; \tag{225}$$

$$V_c = h V_a = \frac{h E_{a1} Z}{Z_1 + Z}. \tag{226}$$

Exercises

1. A 100 MW 0·8 p.f. three-phase a.c. generator generates at 13·8 kV. The positive- and negative-sequence reactances of the machine are 16% and 12% respectively. A fault occurs across phases *a* and *b*. If the resistance of the fault is 100 Ω determine the fault current and the power dissipated in the fault.

2. A 3-phase star-connected alternator, with the star point solidly earthed, is short-circuited (a) between one line terminal and earth, and (b) by two line terminals being simultaneously connected to earth.

Derive for each case an expression for the sustained short circuit current in terms of the alternator voltage and the symmetrical component impedances of the machine. [I.E.E., 1959]

3. A star-connected generator with isolated neutral supplies an unbalanced three-phase load. The generator induced e.m.f.s are balanced and the e.m.f. in the red phase is $E_R = (400 + j0)$ V. The phase sequence is *RYB*. The positive- and negative-sequence impedances per phase of the machine are $j0{\cdot}6\ \Omega$ and $j0{\cdot}4\ \Omega$, respectively. The line currents in the *R* and *Y* lines are $I_R = 50\,(3 - j\sqrt{3})$ A and $I_Y = -50\,(1 + j\sqrt{3})$ A, respectively. Calculate the positive- and negative-sequence currents in the red phase and the potential difference between the *R* and *Y* terminals of the machine.

[I.E.E., 1965]

CHAPTER 6

Equivalent Circuits for Unbalanced Faults

EXPRESSIONS have been derived in the previous chapter for the symmetrical components of current which occur when an unbalanced fault takes place on a three-phase generator. Using these expressions it is now possible to derive an equivalent circuit containing the three impedances Z_0, Z_1, and Z_2 and the positive phase sequence e.m.f. E_{a1}, connected in such a manner that the symmetrical components of current may be shown to flow in the circuit. This then gives a characteristic circuit for the particular type of fault. In drawing these characteristic circuits the points F_0, F_1 and F_2 represent the points of fault corresponding to each of the sequence impedances Z_0, Z_1, and Z_2 respectively, and the points N represent their neutral connections. This idea will then be developed further, in the next chapter, to demonstrate how it is possible to draw characteristic circuits to satisfy the conditions of a fault occurring on a complete power network.

54. Line-to-line fault

(a) *Without impedance in the fault*

It has been shown in § 46 that in the event of a line-to-line fault on an a.c. generator the symmetrical components of current are given by eqns. (142), (149), and (150) as

$$I_{a0} = 0,$$

$$I_{a1} = \frac{E_{a1}}{Z_1 + Z_2},$$

$$I_{a2} = \frac{-E_{a1}}{Z_1 + Z_2}.$$

Also from eqn. (151),

$$V_{a1} = V_{a2} = -I_{a2}Z_2,$$
$$= \frac{Z_2}{Z_1 + Z_2} E_{a1}.$$

Since both I_{a0} and V_{a0} are zero, the zero-sequence impedance Z_0 need not be considered and the only conditions to be satisfied in the characteristic circuit are those for the positive- and negative-phase sequence currents and voltages. If the circuit is connected as shown in Fig. 59, and the current directions of I_{a1} and I_{a2} are as shown, then all the above conditions are satisfied and the circuit represents the characteristic circuit for a line-to-line fault.

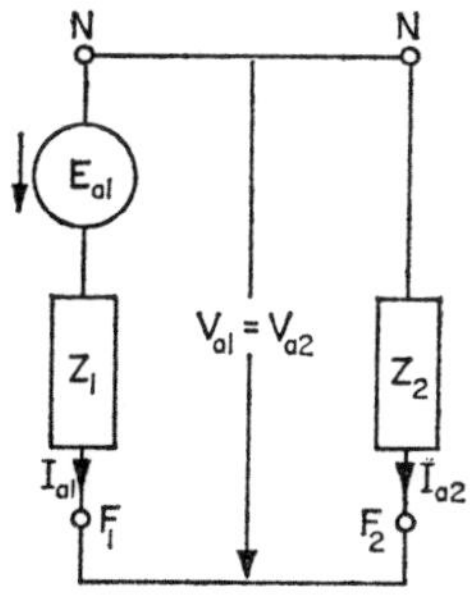

FIG. 59. Equivalent circuit for a line-to-line fault.

(b) *With impedance Z in the fault*

Here, as in the previous case, both I_{a0} and V_{a0} are zero and so, again, Z_0 does not enter into the characteristic circuit.

From eqns. (156) and (157) the symmetrical components of current are given as

$$I_{a1} = \frac{E_{a1}}{Z_1 + Z_2 + Z}$$

and
$$I_{a2} = -I_{a1} = \frac{-E_{a1}}{Z_1 + Z_2 + Z}.$$

For the symmetrical components of voltage we have from eqns. (160) and (161)

$$V_{a1} = \frac{(Z_2 + Z)}{Z_1 + Z_2 + Z} E_{a1},$$

$$V_{a2} = \frac{Z_2 E_{a1}}{Z_1 + Z_2 + Z},$$

and also from eqn. (155) is obtained

$$V_{a1} - V_{a2} = I_{a1} Z.$$

All the above conditions may be satisfied by the characteristic circuit shown in Fig. 60.

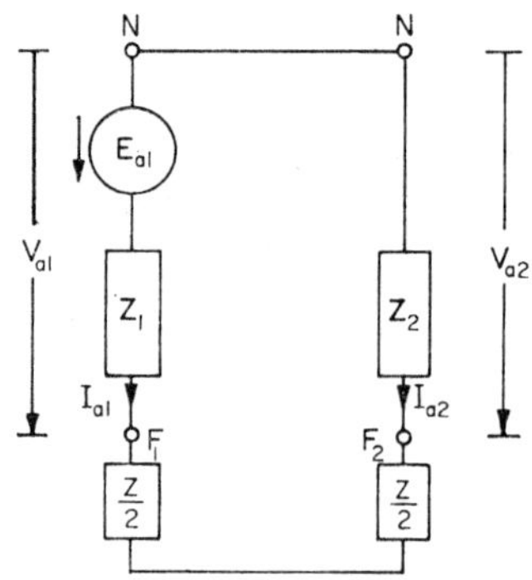

FIG. 60. Equivalent circuit for a line-to-line fault through an impedance.

55. Single line-to-earth fault

(a) *Without impedance in the fault*

From eqn. (169) the symmetrical components of current are given by

$$I_{a0} = I_{a1} = I_{a2} = \frac{E_{a1}}{Z_0 + Z_1 + Z_2}.$$

This time the zero-phase sequence component of current exists, and from the above relationship it is seen that the zero-sequence impedance Z_0 must be included in the characteristic circuit.

Equations (170)–(172) give the symmetrical components of voltage in terms of the known quantities where

$$V_{a0} = \frac{-E_{a1}Z_0}{Z_1 + Z_2 + Z_0} = -I_{a0}Z_0,$$

$$V_{a1} = \frac{E_{a1}(Z_0 + Z_2)}{Z_1 + Z_2 + Z_0} = E_{a1} - I_{a1}Z_1,$$

$$V_{a2} = \frac{-E_{a1}Z_2}{Z_1 + Z_2 + Z_0} = -I_{a2}Z_2.$$

Also $$V_{a0} + V_{a1} + V_{a2} = 0.$$

The symmetrical components of current are equal and the sum of the symmetrical components of voltage at the fault is zero. This gives rise to the characteristic circuit shown in Fig. 61 which satisfies all the above relationships, with the three sequence impedances connected in series.

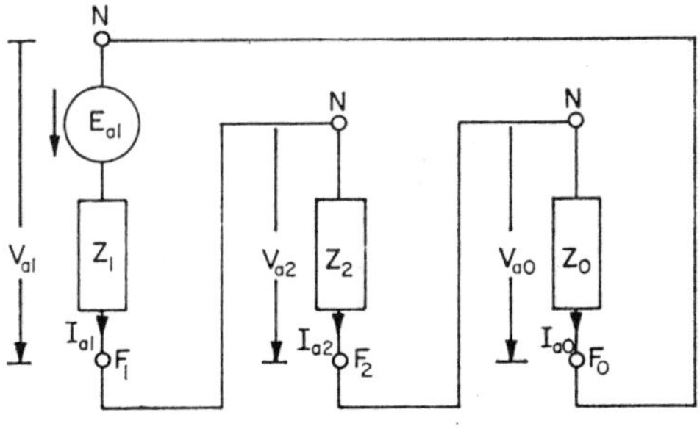

FIG. 61. Equivalent circuit for a line-to-earth fault.

(b) *With impedance Z in the fault*

As in the case of a line-to-earth fault without an impedance in the fault, the three symmetrical components of current are equal and are given in equation (176) by

$$I_{a0} = I_{a1} = I_{a2} = \frac{E_{a1}}{Z_0 + Z_1 + Z_2 + 3Z}.$$

Due to the fault impedance an extra impedance of value $3Z$ has now been included. The symmetrical components of voltage are given by equations (177)–(179) as

$$V_{a0} = \frac{-E_{a1}Z_0}{Z_0 + Z_1 + Z_2 + 3Z} = -I_{a0}Z_0,$$

$$V_{a1} = \frac{E_{a1}(Z_0 + Z_2 + 3Z)}{Z_0 + Z_1 + Z_2 + 3Z} = E_{a1} - I_{a1}Z_1,$$

$$V_{a2} = \frac{-E_{a1}Z_2}{Z_0 + Z_1 + Z_2 + 3Z} = -I_{a2}Z_2.$$

The above relationships are satisfied in the characteristic circuit of Fig. 62, where the three sequence impedances are connected in series with an impedance of three times the fault impedance.

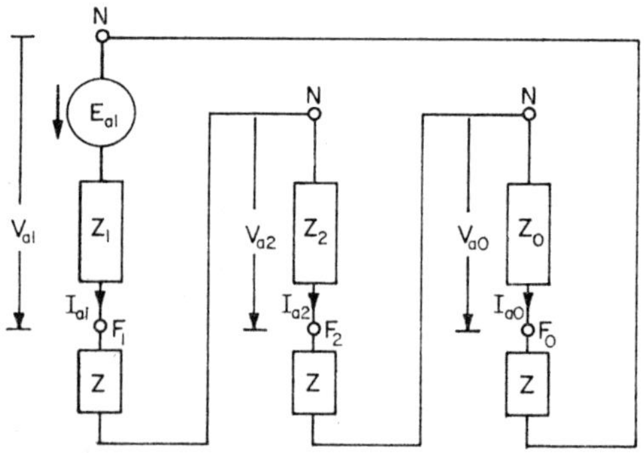

FIG. 62. Equivalent circuit for a line-to-earth fault through an impedance.

56. Double line-to-earth fault

(a) *Without impedance in the fault*

In the case of a double line-to-earth fault the symmetrical components of current are given by equations (187)–(189) where

$$I_{a0} = \frac{-E_{a1}}{Z_1 + (Z_0 Z_2)/(Z_0 + Z_2)} \cdot \frac{Z_2}{Z_0 + Z_2},$$

$$I_{a1} = \frac{E_{a1}}{Z_1 + (Z_0 Z_2)/(Z_0 + Z_2)},$$

$$I_{a2} = \frac{-E_{a1}}{Z_1 + (Z_0 Z_2)/(Z_0 + Z_2)} \cdot \frac{Z_0}{Z_0 + Z_2}.$$

Also the symmetrical components of voltage at the fault, from eqn. (193), are seen to be equal, and are given by

$$V_{a0} = V_{a1} = V_{a2} = \frac{E_{a1}}{Z_1 + (Z_0 Z_2)/(Z_0 + Z_2)} \cdot \frac{Z_0 Z_2}{Z_0 + Z_2}.$$

The impedance term in the expression for I_{a1} suggests a connection of impedances such that Z_1 is connected in series with a parallel connection of Z_0 and Z_2. Connecting this arrangement of impedances with the voltage E_{a1} gives the characteristic circuit shown in Fig. 63 which satisfies all the above relationships for the symmetrical components of current and voltage at the fault.

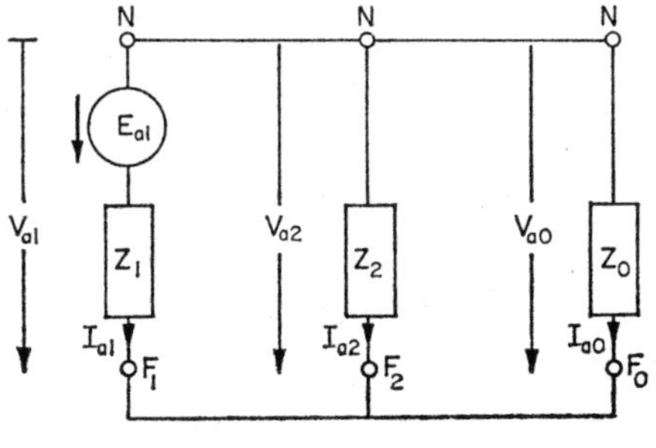

FIG. 63. Equivalent circuit for a double line-to-earth fault.

(b) *With impedance Z in the fault*

The symmetrical components of current in this case are given in eqns. (200)–(202) where

$$I_{a0} = \frac{-E_{a1}}{Z_1 + [Z_2(Z_0 + 3Z)]/(Z_0 + Z_2 + 3Z)} \cdot \frac{Z_2}{Z_0 + Z_2 + 3Z},$$

$$I_{a1} = \frac{E_{a1}}{Z_1 + [Z_2(Z_0 + 3Z)]/(Z_0 + Z_2 + 3Z)},$$

$$I_{a2} = \frac{-E_{a1}}{Z_1 + [Z_2(Z_0 + 3Z)]/(Z_0 + Z_2 + 3Z)} \cdot \frac{Z_0 + 3Z}{Z_0 + Z_2 + 3Z}.$$

The expressions for the symmetrical components of voltage at the fault are given in eqns. (206) and (207) by

$$V_{a0} = \frac{E_{a1}}{Z_1 + [Z_2(Z_0 + 3Z)]/(Z_0 + Z_2 + 3Z)} \cdot \frac{Z_0 Z_2}{Z_0 + Z_2 + 3Z}$$

and

$$V_{a1} = V_{a2} = \frac{E_{a1}}{Z_1 + [Z_2(Z_0 + 3Z)]/(Z_0 + Z_2 + 3Z)} \cdot \frac{Z_2(Z_0 + 3Z)}{Z_0 + Z_2 + 3Z}.$$

Comparing the impedance terms in the expressions for the symmetrical components of current for cases (a) and (b) with a double line-to-earth fault it is seen that in case (b) the zero sequence impedance Z_0 has an extra term $3Z$ added to it throughout. The characteristic circuit for case (b) is then seen to be the same as in case (a) except that the impedance Z_0 has an extra impedance equal to three times the fault impedance connected in series with it. This circuit is shown in Fig. 64 and it can be seen to satisfy the relationships for the symmetrical components of both current and voltage.

57. Three-phase fault

(a) *Without impedance in the fault*

With a three-phase fault the system remains balanced and only the positive-phase sequence current and voltage need be con-

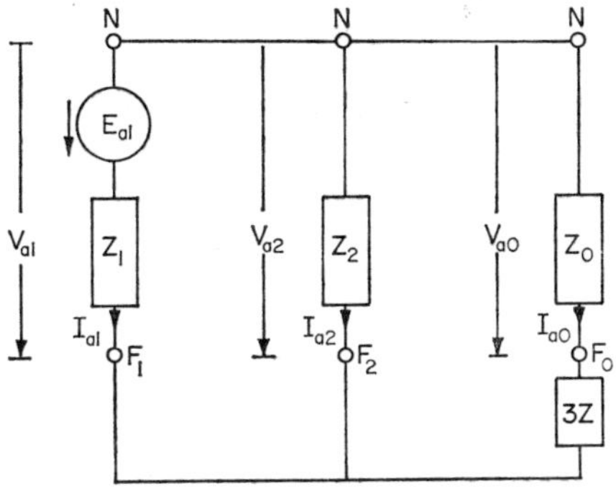

FIG. 64. Equivalent circuit for a double line-to-earth fault through an impedance.

sidered. The positive-phase sequence current is given in eqn. (212) by

$$I_{a1} = \frac{E_{a1}}{Z_1}$$

and the positive-phase sequence voltage V_{a1} is zero. Thus the characteristic circuit is represented by the e.m.f. E_{a1} applied across the positive sequence impedance Z_1 as in Fig. 65.

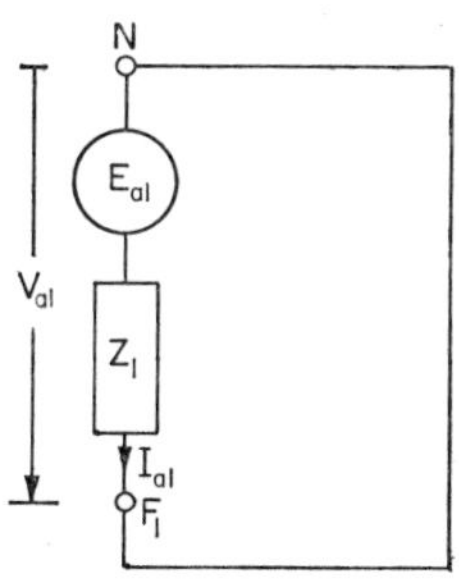

FIG. 65. Equivalent circuit for a three-phase fault.

(b) *With impedance Z in the fault*

Again if an impedance Z, which is assumed equal in each of the three phases, occurs in the fault the system will be balanced and only the positive phase sequence components of current and voltage are to be considered.

Equation (217) gives the positive phase-sequence current as

$$I_{a1} = \frac{E_{a1}}{Z_1 + Z}$$

and from eqn. (222) the positive phase sequence voltage at the fault is

$$V_{a1} = I_{a1}Z.$$

The characteristic circuit as shown in Fig. 66 fulfils both of these conditions and consists of the e.m.f. E_{a1} applied across Z_1 and Z in series.

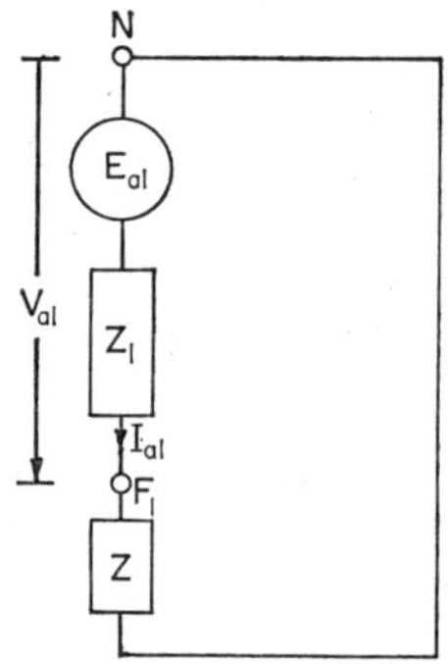

FIG. 66. Equivalent circuit for a three-phase fault through an impedance.

Exercise

1. In the case of a three-phase generator having an earthed neutral, and simultaneous short circuits occurring across phases b and c, and from phase a to earth, show that, if the generator impedances are Z_0, Z_1, and Z_2 and the internal voltage is E_{a1}, the equivalent circuit is given in Fig. 67.

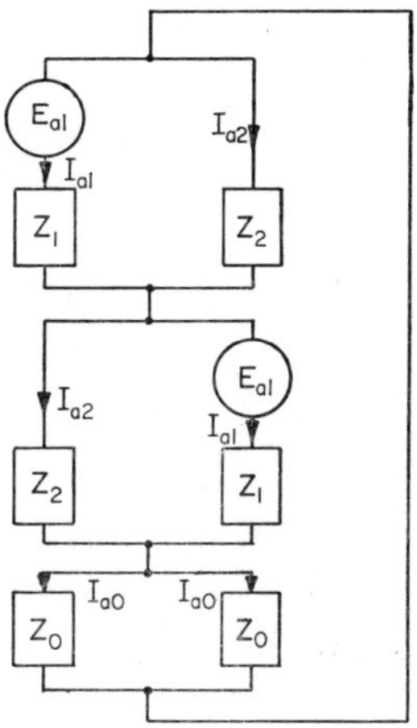

FIG. 67. Exercise 1.

CHAPTER 7

Sequence Networks and Faults on Three-phase Systems

IT WILL be shown that, just as in the simple case of a three-phase a.c. generator where the unsymmetrical fault condition may be resolved by means of an equivalent circuit consisting of a suitable connection of its sequence impedances, the same may be done for a complete balanced three-phase system having an unsymmetrical fault. Here the situation is somewhat more complex since whereas with a single generator only three impedances (the positive-, negative-, and zero-sequence impedances) may be involved, in the case of a three-phase system three complete networks, the positive-, negative-, and zero-sequence networks, must first be derived before making suitable connections of these networks in order to calculate the required fault conditions.

Throughout the chapter it is assumed that the three-phase systems considered are balanced, which, in general, will be the case.

58. Unbalanced fault on a three-phase system

What we now consider is really an extension of Thevenin's theorem applied to a balanced three-phase network. Consider an active network as shown by means of a block diagram in Fig. 68. This may be used to represent a complete power network. At a given point in the network, connections are made to the three phases, and also to the neutral, and brought out at the terminals a, b, c, and N.

As the network is assumed to be balanced, the phase voltages between a and N, b and N, and c and N will represent a balanced system of three-phase voltages, and if V_{aN} is assumed to equal E, then

$$V_{bN} = h^2E \quad \text{and} \quad V_{cN} = hE.$$

The value of E will depend upon the active voltages in the network, their method of connection to one another, and also the arrangement of the impedances contained in the network.

If now the network is made passive by short-circuiting all the internal e.m.f.s and replacing them by their internal impedances, the resulting network, consisting only of impedances, will be balanced. The three values of impedance measured between a and

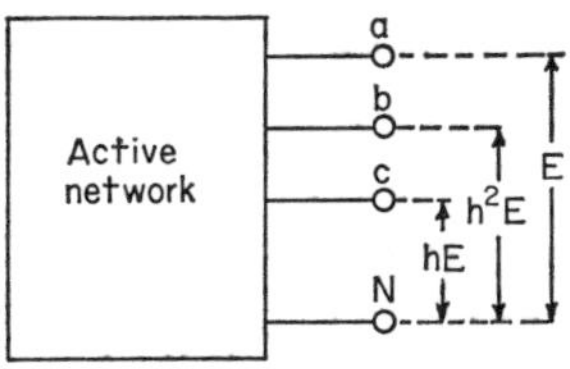

FIG. 68. Active power network.

N, b and N, and c and N will thus be equal. Let the value of this impedance equal Z. The network is now equivalent to a star-connected system of e.m.f. E per phase and impedance Z per phase, as shown in Fig. 69, where Z is a combination of all the impedances in the network as viewed between the terminals a and N. The currents flowing in any load connected to the terminals a, b, c, and N of the original active network of Fig. 68 may now be solved by connecting the same load to the terminals a, b, c, and N in the equivalent network of Fig. 69 which is seen to be equivalent to a three-phase a.c. generator. Taking this argument one step further, it may now be seen that, for an unbalanced fault condition on a three-phase network, the problem may be resolved by reducing the original network to an equivalent circuit of a single three-phase machine and applying the fault to this equivalent circuit. It should be noted that, whereas in the case of an unbalanced fault on the three-phase a.c. generator only three impedances of positive-, negative-, and zero-sequence impedance are necessary, in the case of a complete power network each of these single impedances will now be replaced by a complete sequence network to be derived from the original power network. Each of the sequence networks may then be simplified and finally reduced to a single equivalent sequence impedance.

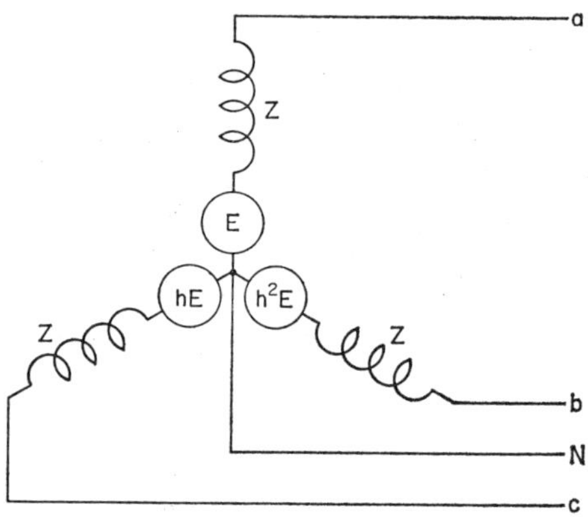

FIG. 69. Equivalent network.

In finding the solution for a particular type of fault on a complete power system, use will be made of the characteristic circuits developed in Chapter 6, except that each of the sequence impedances will now be replaced by a sequence impedance network. The terminals *a*, *b*, *c*, and *N* of Fig. 68 will be brought out at the point of fault on this system.

Just as when considering an unbalanced fault on a three-phase machine, it will only be necessary to consider one phase of a three-phase network which may be drawn as a one-line diagram. From this the three sequence networks may be derived. Each of the sequence networks will then have its own neutral point, line conductors, and point of fault.

59. Positive-sequence network

The positive-sequence network will be the same as the normal active network under consideration. It will contain the e.m.f.s of any generators in the original network and the appropriate values of positive-sequence impedance for the generators and other components in the network. This network will carry the appropriate values of positive-phase sequence current in each of its branches.

60. Negative-sequence network

In general the negative-sequence network will be similar to the positive-sequence network, except that it will contain the appropriate values of negative-sequence impedance for its components, and will normally be a passive network containing no e.m.f.s since negative-phase sequence e.m.f.s will not usually be generated. It will contain the same number of branches as the positive-sequence network and will carry only negative-phase sequence currents.

61. Zero-sequence network

The zero-sequence network will again be a passive network since normally e.m.f.s of zero-phase sequence will not be generated. It will usually be somewhat different to the positive- and negative-sequence networks in that it will contain fewer branches. The values to be taken for zero-sequence impedance of generators, transformers, and three-phase loads will be dependent upon whether they are delta- or star-connected and if star-connected, whether they are earthed or not. Special consideration will be given to zero-sequence impedances in subsequent sections.

62. Zero-sequence impedances

Of the three sequence networks, the positive- and negative-sequence networks are generally fairly straightforward and are similar in form to the original network when drawn as a single line diagram except that the negative-sequence network will not normally contain any e.m.f.s. It is the zero-sequence network which may cause most trouble in its derivation since there are certain special requirements to be considered in its construction; however, it should not normally involve any great difficulty provided certain rules are observed. The important requirement for the zero-sequence impedance to have a finite value is that a fault path must be available via earth, and in the case of transformers it is also necessary that a zero-phase sequence current path must exist on two windings.

(a) *Series and shunt impedances*

Figure 70 shows the circuit diagrams and equivalent zero-sequence networks for various connections of both series and shunt impedances.

(*I*) *Series impedance without neutral conductor.* Since there is no neutral conductor there can be no zero-phase sequence currents in the series impedances as there is no return path for the currents, and so the zero-sequence impedance must be represented by an impedance of infinite value.

(*II*) *Series impedance with neutral conductor.* Here the zero-phase sequence currents will have a return path through the neutral conductor, and so in a single line diagram the zero-sequence impedance will be represented by a straight-through impedance of value Z.

(*III*) *Unearthed shunt impedance.* As in case (I), since there is no neutral connection, there will be no return path for the zero-phase sequence currents and so, again, the zero-sequence impedance will be infinite.

(*IV*) *Earthed shunt impedance.* The zero-phase sequence

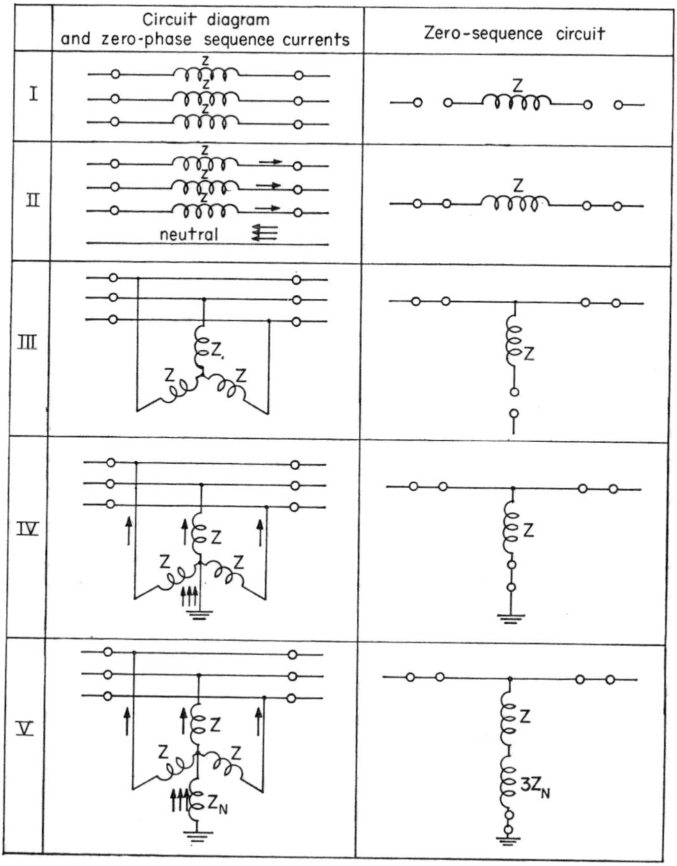

FIG. 70. Equivalent zero-phase sequence circuits for series and shunt impedances.

currents will have a return path through the earth connection giving a zero-sequence impedance to earth of Z as shown.

(*V*) *Shunt impedance earthed through a neutral impedance* Z_N. A return path for the zero-phase sequence current is here provided through the neutral earthing impedance Z_N. As the zero-phase sequence currents I_{a0}, I_{b0}, and I_{c0} flowing in the three phases are equivalent to a current of $3I_{a0}$ flowing in the neutral and a voltage drop of $3I_{a0}Z_N$ in the neutral impedance, the equivalent zero-sequence impedance of the neutral impedance must be equal to $3Z_N$. This impedance will be connected in series with an impedance of value Z in the equivalent zero-sequence network to give the total zero-sequence impedance to earth.

(b) *Two-winding transformers*

The equivalent circuit of a two-winding transformer may be represented by a T-network consisting of two impedances connected in series, representing the primary and secondary winding leakage impedances, and a shunt impedance, representing the exciting impedance. Since the exciting impedance is very large compared with the primary and secondary leakage impedances, so far as short-circuit calculations are concerned, it may be regarded as being infinite.

The positive- and negative-sequence impedances of a transformer are equal, and are given by the series impedance of the T-network and so are represented in the equivalent circuit as straight-through impedances.

In order to determine the zero-sequence impedance the following two considerations must be made for each winding:

(1) Whether zero-phase sequence currents can flow into the winding being considered from the external circuit. This will only be so when the winding is star-connected and earthed at its neutral point, either directly or indirectly, through a neutral impedance. If there is a circuit for zero-phase sequence currents then this winding is shown connected to the external circuit; if not, as in the case of an

unearthed star- or delta-connected winding, then it is shown disconnected from the external circuit.

(2) Whether zero-phase sequence currents are able to flow in the winding alone without flowing in the external circuit as in the case of a delta-connected winding. If they are, this winding is shown connected to earth, but disconnected from the external circuit.

In Fig. 71 are shown the equivalent zero-sequence circuits for various types of two-winding transformers.

(*I*) *Unearthed star–star*. Since there is no earth connection on either winding, no path exists for the zero-phase sequence currents to flow external to the transformer circuit. Both windings must therefore be represented disconnected from the external circuit. Also, as both windings are star-connected, zero-phase sequence currents cannot circulate within the windings either. They are both thus shown disconnected from earth. Such an arrangement of windings therefore presents an infinite impedance to the flow of zero-phase sequence currents.

(*II*) *Unearthed star–earthed star*. There is no path for zero-phase sequence currents for an earth fault on the primary winding, hence this winding is shown disconnected from the external circuit in the zero-sequence circuit. For the secondary winding this is also virtually true in the case of three separate single phase transformers, or a three-phase shell type transformer, since there would be no path for primary zero-phase sequence currents. However, with a three-phase core-type transformer, it is possible for the fluxes produced by the zero-phase sequence currents in the secondary winding to find high reluctance paths outside the core and in such parts as the tank walls. Since these paths have a relatively high reluctance it may be possible for the zero-sequence reactance to have a value as low as 0·3 per unit.

(*III*) *Earthed star–star*. Zero-phase sequence currents are able to flow into both windings from the external circuit since both star points are earthed. The transformer may thus be represented in the zero-sequence circuit by a straight-through impedance.

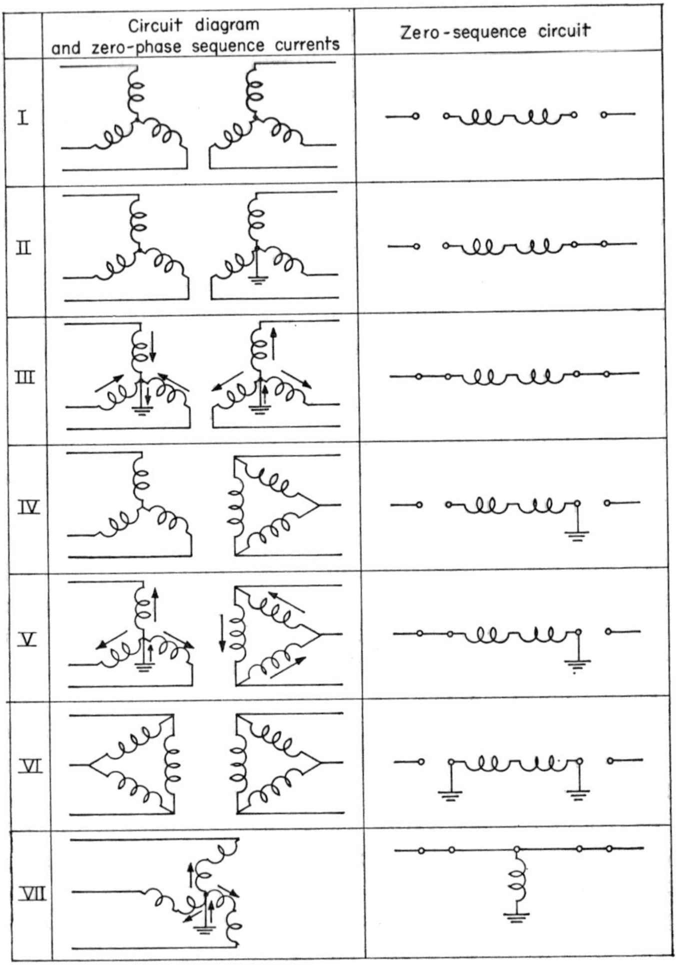

FIG. 71. Equivalent zero-phase sequence circuits for two-winding transformers.

(*IV*) *Unearthed star–delta.* Since zero-phase sequence currents are unable to flow from the external circuit to either the primary or secondary winding they are both shown open-circuited in the zero-sequence circuit.

(*V*) *Earthed star–delta.* As the star-connected winding is earthed, current of zero-phase sequence may circulate in the primary, and corresponding currents will flow in the delta winding which is shown connected to earth in the zero-sequence circuit.

(*VI*) *Delta–delta.* Both windings must be represented as disconnected from the external circuit as no zero-phase sequence currents are able to flow into or out of either winding.

(*VII*) *Earthing transformer.* The earthing transformer has two separate windings on one limb which are represented in the circuit diagram by parallel branches. Neglecting the magnetizing currents it will have infinite impedance for positive- and negative-phase sequence currents. For zero-phase sequence currents the ampere turns due to each winding on the same limb cancel, and the impedance becomes the leakage impedance of the windings. The zero-sequence impedance may thus be represented by an impedance to earth.

(c) *Three-winding transformers*

The same consideration must be applied to three-winding transformers as in the case of two-winding transformers in determining the zero-sequence circuit and must be applied to each of the three windings in turn. Figure 72 shows the zero-sequence circuits for three-winding transformers having delta-connected tertiary windings.

(*I*) *Unearthed star–star.* As for the two-winding unearthed star–star transformer, no paths exist for zero-phase sequence currents to flow between either the primary or secondary windings and the external circuit and so the zero-sequence impedance is infinite.

(*II*) *Unearthed star–earthed star.* Zero-phase sequence currents can flow between the external circuit and the secondary winding.

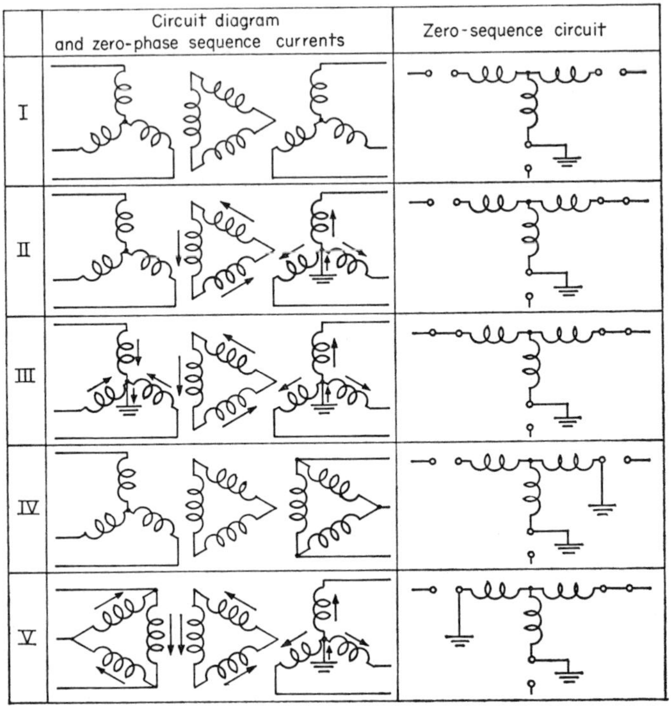

FIG. 72. Equivalent zero-phase sequence circuits for three-winding transformers.

This will give rise to currents in the tertiary but not in the primary winding since it has no earth connection.

(*III*) *Earthed star–star.* All three windings can carry zero-phase sequence current; the zero-sequence circuit being represented by two straight-through impedances connected in series, representing the primary and secondary windings, with the tertiary winding represented by an impedance connected to earth.

(*IV*) *Unearthed star–delta.* Since the primary winding is unearthed and the secondary is connected in delta, zero-phase sequence currents are unable to flow from the external circuit and so the zero-sequence impedance will be infinite.

(*V*) *Delta–earthed star*. Zero-phase sequence currents are able to flow between the external circuit and the secondary winding only. This will give rise to currents in both the primary and tertiary windings since they are both connected in delta.

63. Method of fault calculation for a network

The following procedure should be adopted in determining the fault conditions when an unbalanced fault occurs on a three-phase network:

(1) Represent the network as a single-line diagram.
(2) Construct the positive-sequence network inserting the appropriate values of positive-sequence impedance, and the positive-sequence e.m.f.s due to generators.
(3) Reduce the network of impedances in the positive-sequence network to a single equivalent impedance.
(4) Construct the negative-sequence network inserting the appropriate values of negative-sequence impedance.
(5) Reduce the negative-sequence network to a single equivalent impedance.
(6) In the case of a fault involving earth construct the zero-sequence network.
(7) Reduce the zero-sequence network to a single equivalent impedance.
(8) Connect the equivalent positive-, negative-, and zero-sequence impedances in the form of the appropriate characteristic circuit for the particular type of fault under consideration.
(9) Calculate the phase-sequence components of current at the fault.
(10) It is now possible to determine the phase-sequence components of current at each point in the network and hence the actual values of fault current if required. Also the values

of voltage at various points in the network may be calculated if necessary.

The method may be demonstrated by means of a simple example.

EXAMPLE 13. Two 20 MVA, 6·6 kV synchronous generators are connected in parallel, and supply a 6·6 kV feeder. One generator has its star point earthed through a resistor of 0·5 Ω and the other has its star point isolated.

If an earth fault occurs on one phase at the far end of the feeder determine:

(a) The fault current.
(b) The voltage to earth of the neutral point of the earthed generator.
(c) The voltages to earth of the sound phases at the point of fault.

The impedances are:

	Generators Per unit	Feeder Ω/phase
To positive sequence currents	$j0 \cdot 15$	$j0 \cdot 3$
To negative sequence currents	$j0 \cdot 10$	$j0 \cdot 3$
To zero sequence currents	$j0 \cdot 05$	$j0 \cdot 6$

[I.E.E., May 1962]

The generator line voltage is 6·6 kV.

$$\text{Therefore phase voltage} = \frac{6{,}600}{\sqrt{3}} = 3810 \text{ V}.$$

$$\text{Full-load current of each generator} = \frac{20 \times 10^6}{3 \times 3810} = 1750 \text{ A}.$$

Since it is convenient to work in terms of per unit quantities, it is necessary to convert the ohmic values of feeder and earthing resistor impedance to per unit values.

$$Z_1 \text{ for feeder} = \frac{j0\cdot 3 \times 20}{6\cdot 6^2} = j0\cdot 138 \text{ p.u.}$$

$$Z_2 \text{ for feeder} = \frac{j0\cdot 3 \times 20}{6\cdot 6^2} = j0\cdot 138 \text{ p.u.}$$

$$Z_0 \text{ for feeder} = \frac{j0\cdot 6 \times 20}{6\cdot 6^2} = j0\cdot 276 \text{ p.u.}$$

Let Z_n be the value of the earthing resistor in per unit then

$$Z_n = \frac{0\cdot 5 \times 20}{6\cdot 6^2} = 0\cdot 229 \text{ p.u.}$$

Figure 73a shows the network represented as a single-line diagram and in Fig. 73b the positive-, negative-, and zero-sequence networks have been drawn and reduced to single equivalent impedances.

(a) The characteristic circuit for a line-to-earth fault is as shown in Fig. 73c and from eqn. (176) the phase sequence components of current are given by

$$I_{a0} = I_{a1} = I_{a2} = \frac{E_{a1}}{Z_1 + Z_2 + Z_0 + 3Z_n}.$$

Assuming both generators to have been generating at normal voltage prior to the fault, then

$$E_{a1} = 1 \text{ p.u.}$$

Therefore

$$I_{a0} = I_{a1} = I_{a2} = \frac{1}{j0\cdot 213 + j0\cdot 188 + j0\cdot 326 + 0\cdot 687},$$

$$= \frac{1}{0\cdot 687 + j0\cdot 727},$$

$$= 0\cdot 686 - j0\cdot 725 \text{ p.u.}$$

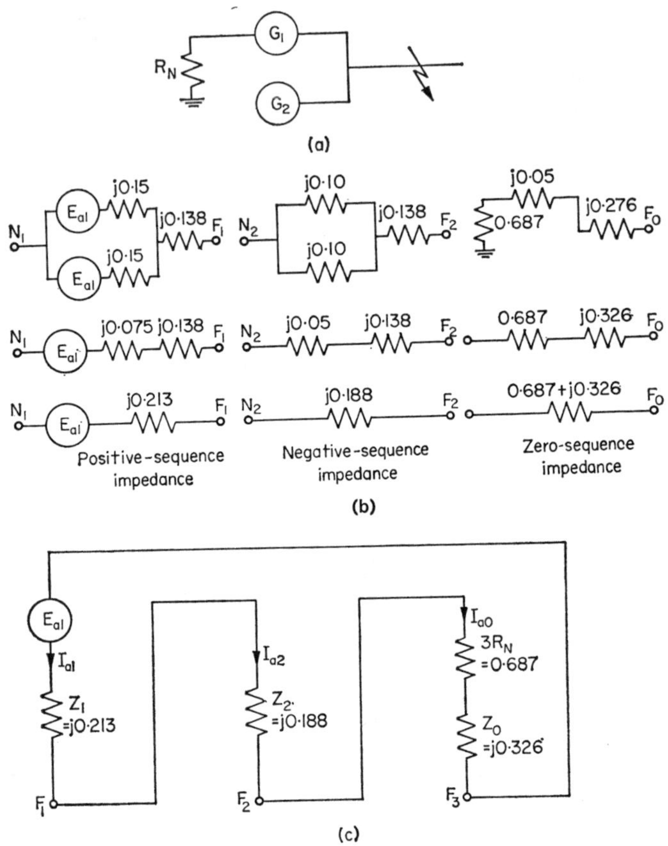

FIG. 73. Example 13—Sequence impedances.

The fault current in phase a is given by

$$I_a = I_{a0} + I_{a1} + I_{a2},$$

$$= 3(0\cdot686 - j0\cdot725),$$

$$= 2\cdot058 - j2\cdot175 \text{ p.u.}$$

Therefore
$$|I_a| = \sqrt{(2\cdot058^2 + 2\cdot175^2)},$$
$$= 2\cdot99 \text{ p.u.},$$
$$\equiv 1750 \times 2\cdot99,$$
$$= 5230 \text{ A}.$$

(b) The voltage to earth of the neutral point is given from eqn. (74) by

$$V_n = -I_{a0}(3Z_n),$$
$$= -(0\cdot686 - j0\cdot725)(3 \times 0\cdot229),$$
$$= -0\cdot472 + j0\cdot498 \text{ p.u.}$$

Therefore
$$|V_n| = \sqrt{(0\cdot472^2 + 0\cdot498^2)},$$
$$= 0\cdot686 \text{ p.u.},$$
$$\equiv 3810 \times 0\cdot686,$$
$$= 2615 \text{ V}.$$

(c) The per unit values of the phase-sequence components of voltage at the fault are given by:

$$V_{a1} = E_{a1} - I_{a1}Z_1,$$
$$= 1 - (0\cdot686 - j0\cdot725)\,j0\cdot213,$$
$$= 1 - 0\cdot155 - j0\cdot146,$$
$$= 0\cdot845 - j0\cdot146;$$
$$V_{a2} = -I_{a2}Z_2,$$
$$= -(0\cdot686 - j0\cdot725)\,j0\cdot188,$$
$$= -0\cdot136 - j0\cdot129;$$
$$V_{a0} = -I_{a0}\,(Z_0 + 3Z_n),$$
$$= -(0\cdot686 - j0\cdot725)\,(0\cdot687 + j0\cdot326),$$
$$= -0\cdot709 + j0\cdot276.$$

Using the expressions given by eqn. (63) we have

$$\begin{aligned}
V_b &= V_{a0} - \tfrac{1}{2}(V_{a1} + V_{a2}) - j\tfrac{1}{2}\sqrt{(3)}\,(V_{a1} - V_{a2}),\\
&= -\,0\cdot709 + j0\cdot276 - \tfrac{1}{2}(0\cdot845 - j0\cdot146\\
&\quad - 0\cdot136 - j0\cdot129) - j\tfrac{1}{2}\sqrt{(3)}\,(0\cdot845\\
&\quad - j0\cdot146 + 0\cdot136 + j0\cdot129),\\
&= -\,0\cdot709 + j0\cdot276 - \tfrac{1}{2}(0\cdot709 - j0\cdot275)\\
&\quad - j\tfrac{1}{2}\sqrt{(3)}\,(0\cdot981 - j0\cdot017),\\
&= -\,1\cdot079 - j0\cdot436 \text{ p.u.}
\end{aligned}$$

$$\begin{aligned}
\text{Therefore } |\,V_b\,| &= \sqrt{(1\cdot079^2 + 0\cdot436^2)},\\
&= 1\cdot162 \text{ p.u.},\\
&\equiv 3810 \times 1\cdot162 \text{ V},\\
&= 4430 \text{ V};
\end{aligned}$$

$$\begin{aligned}
V_c &= V_{a0} - \tfrac{1}{2}(V_{a1} + V_{a2}) + j\tfrac{1}{2}\sqrt{(3)}\,(V_{a1} - V_{a2}),\\
&= -\,0\cdot709 + j0\cdot276 - \tfrac{1}{2}(0\cdot709 - j0\cdot275)\\
&\quad + j\tfrac{1}{2}\sqrt{(3)}\,(0\cdot981 - j0\cdot017),\\
&= -\,1\cdot049 + j1\cdot264 \text{ p.u.}
\end{aligned}$$

$$\begin{aligned}
\text{Therefore } |\,V_c\,| &= \sqrt{(1\cdot049^2 + 1\cdot264^2)},\\
&= 1\cdot64 \text{ p.u.},\\
&\equiv 3810 \times 1\cdot64 \text{ V},\\
&= 6250 \text{ V}.
\end{aligned}$$

Although not asked for in the above question, it is also possible to find the current distribution throughout the network. In Fig. 74a is shown the distribution of the phase sequence components of current in the sequence networks. The phase sequence currents have then been added as in equations (70) to give the current distribution in each phase as shown in Fig. 74b where each phase is

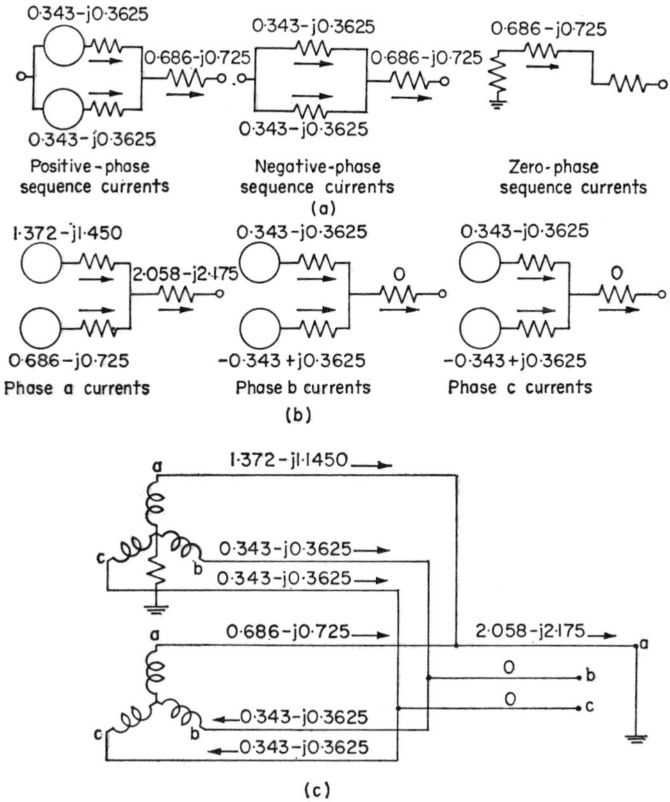

FIG. 74. Example 13—Current distribution.

represented as a single line diagram. Figure 74c shows the complete three-phase network and the fault current distribution in terms of per unit current.

64. Use of analogues

Whilst the previous example illustrates how fault conditions may be solved by calculation for a very simple network the amount of work involved in solving an elaborate transmission system involving three or more interconnected generators would be considerable. For this reason it is desirable to resort to other methods. One such method is to construct a model or analogue of the system and relate measurements made on the analogue to the actual network.

Adopting the analogue method the sequence networks are derived as for calculation. Models of the networks are then constructed with suitable resistors, reactors, and capacitors. The values used in the models may, if convenient, represent the true values to some suitable scale. The network e.m.f.s may be supplied to scale by means of a phase-shifting transformer. The three-sequence networks are then connected together with any fault impedance to represent the actual fault conditions. With all the voltages and circuit components set at their scaled values, readings to scale may now be made of fault conditions using suitable instruments at any point in the system. In order to measure the phase currents of phase *a* instead of the phase-sequence currents the primaries of three-current transformers may be connected in with the three-phase sequence components and the phase current obtained by summation of the three secondaries. All readings obtained must then be multiplied by a suitable scale factor to represent actual values on the system.

In practice an a.c. network analyser would be used for the solution of this type of problem. It consists fundamentally of a very flexible electrical system consisting of a number of variable resistance, reactor, and capacitor components together with a number of single-phase a.c. voltage supplies which are variable in both magnitude and phase. A large range of metering devices is also included and by means of plugboards an extensive arrangement of circuits may be built up to similate any particular network problem, usually by scaling the actual network values to suit the components available on the network analyser.

Exercises

1. Two similar star-connected generators, one of which has its star connected point earthed, supply 11 kV three phase bus-bars. An 11/66 kV delta-star connected transformer, with its star point earthed, is supplied from the bus-bars.

The impedances, referred to 11 kV are as given below:

	Generator	Transformer
Z_1	$j3 \cdot 0$	$j3 \cdot 0$
Z_2	$j2 \cdot 0$	$j3 \cdot 0$
Z_0	$j1 \cdot 0$	$j3 \cdot 0$

Determine the fault current for a single earth fault (a) on the bus-bar (b) on the H.V. terminal of the transformer.

Find the voltage to earth of the two sound bus-bars in case (a).

[I.E.E., 1956]

2. Develop in terms of the generated e.m.f. and the impedance to positive and negative sequence components of current, and with an explanation of each step, an expression for the current and the voltages to earth of each phase if a phase-to-phase fault occurs at the terminals of a three phase generator having its star point earthed.

An 11 kV three phase generator, a transformer, and a 66 kV line, have the following reactances to positive and negative sequence components of current, the resistances being negligible and values being referred to 66 kV.

	Generator	Transformer	Line
Z_1 Ω/phase	$j3$	$j2$	$j1 \cdot 5$
Z_2 Ω/phase	$j2$	$j2$	$j1 \cdot 5$

If a phase-to-phase fault occurs at the far end of the line find:

(a) The fault current.

(b) The negative sequence component of voltage at the H.V. terminals of the transformer. [I.E.E., 1961]

CHAPTER 8

Unbalanced Loads

THE foregoing analysis of the theory of symmetrical components has been based on the assumption of balanced loads, when it has been shown that there is no interaction between voltages and currents of different phase sequences. Symmetrical component theory will now be applied to the case of unbalanced loads when it will be seen that there is interaction between voltages and currents of different phase sequences, and that voltage drops of one phase sequence may be produced by currents of different phase sequence, in addition to currents of the same phase sequence.

Examples of unbalanced three-phase loads are to be found, for instance, in the case of three separate single-phase loads, unsymmetrical three-phase transmission lines, and unsymmetrical transformer banks. The former case will be studied in more detail.

65. Unbalanced star-connected load

An unbalanced star-connected load is represented in Fig. 75 where Z_a, Z_b, and Z_c are three unequal impedances. Applied to the load is a balanced three-phase supply of voltage E_a, E_b, and E_c per phase respectively, and of positive phase sequence. Unbalanced currents I_a, I_b, and I_c flow in the three phases and the unbalanced voltages appearing across the three-load impedances are V_a, V_b, and V_c respectively.

Although this circuit may be solved quite easily by other methods, it has been chosen here as a simple circuit by means of which the method of symmetrical components may be demonstrated for unbalanced loads.

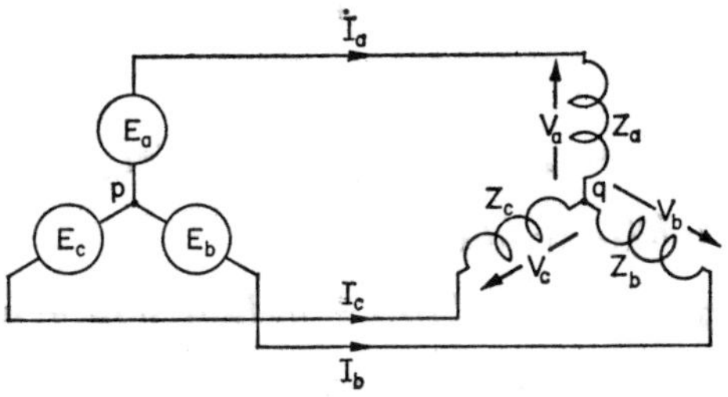

FIG. 75. Unbalanced star-connected load.

Inspection of the circuit shows the phase voltages to be given by

$$\left.\begin{aligned} V_a &= I_a Z_a, \\ V_b &= I_b Z_b, \\ V_c &= I_c Z_c. \end{aligned}\right\} \tag{227}$$

Replacing the phase currents by the phase sequence components of current, we have

$$\left.\begin{aligned} V_a &= (I_{a0} + I_{a1} + I_{a2})Z_a, \\ V_b &= (I_{a0} + h^2 I_{a1} + h I_{a2})Z_b, \\ V_c &= (I_{a0} + h I_{a1} + h^2 I_{a2})Z_c. \end{aligned}\right\} \tag{228}$$

Also expressing the symmetrical components of voltage in terms of the phase voltages

$$\left.\begin{aligned} V_{a0} &= \tfrac{1}{3}(V_a + V_b + V_c), \\ V_{a1} &= \tfrac{1}{3}(V_a + hV_b + h^2 V_c), \\ V_{a2} &= \tfrac{1}{3}(V_a + h^2 V_b + hV_c), \end{aligned}\right\} \tag{229}$$

and substituting the values for the phase voltages from eqns. (228) into eqn. (229) gives

$$V_{a0} = \tfrac{1}{3}[(I_{a0} + I_{a1} + I_{a2})Z_a + (I_{a0} + h^2 I_{a1} + hI_{a2})Z_b + (I_{a0} + hI_{a1} + h^2 I_{a2})Z_c],$$

$$V_{a1} = \tfrac{1}{3}[(I_{a0} + I_{a1} + I_{a2})Z_a + h(I_{a0} + h^2 I_{a1} + hI_{a2})Z_b + h^2(I_{a0} + hI_{a1} + h^2 I_{a2})Z_c],$$

$$V_{a2} = \tfrac{1}{3}[(I_{a0} + I_{a1} + I_{a2})Z_a + h^2(I_{a0} + h^2 I_{a1} + hI_{a2})Z_b + h(I_{a0} + hI_{a1} + h^2 I_{a2})Z_c],$$

Regrouping the right-hand side of the above expressions leads to the result

$$\left.\begin{aligned} V_{a0} &= \tfrac{1}{3}[(Z_a + Z_b + Z_c)I_{a0} + (Z_a + h^2 Z_b + hZ_c)I_{a1} + (Z_a + hZ_b + h^2 Z_c)I_{a2}],\\ V_{a1} &= \tfrac{1}{3}[(Z_a + hZ_b + h^2 Z_c)I_{a0} + (Z_a + Z_b + Z_c)I_{a1} + (Z_a + h^2 Z_b + hZ_c)I_{a2}],\\ V_{a2} &= \tfrac{1}{3}[(Z_a + h^2 Z_b + hZ_c)I_{a0} + (Z_a + hZ_b + h^2 Z_c)I_{a1} + (Z_a + Z_b + Z_c)I_{a2}]. \end{aligned}\right\} \tag{230}$$

Inspection of the above equations shows that there are three impedance terms repeated in each equation. Hence, putting

$$\left.\begin{aligned} \tfrac{1}{3}(Z_a + Z_b + Z_c) &= \zeta_0,\\ \tfrac{1}{3}(Z_a + hZ_b + h^2 Z_c) &= \zeta_1,\\ \tfrac{1}{3}(Z_a + h^2 Z_b + hZ_c) &= \zeta_2, \end{aligned}\right\} \tag{231}$$

we may write eqn. (230) now as

$$\left.\begin{aligned} V_{a0} &= \zeta_0 I_{a0} + \zeta_2 I_{a1} + \zeta_1 I_{a2},\\ V_{a1} &= \zeta_1 I_{a0} + \zeta_0 I_{a1} + \zeta_2 I_{a2},\\ V_{a2} &= \zeta_2 I_{a0} + \zeta_1 I_{a1} + \zeta_0 I_{a2}. \end{aligned}\right\} \tag{232}$$

In eqns. (231) it is important to note that the terms ζ_0, ζ_1, and ζ_2 are not the same as the zero-, positive-, and negative-sequence impedances Z_0, Z_1, and Z_2 as used previously, in the case of balanced loads.

Consideration of eqns. (232) shows that, when the load is unbalanced, the voltage of any one-phase sequence is due to currents of all three-phase sequences; hence there is interaction between the phase sequences. This, of course, is contrary to the results produced with balanced loads, where it was found that interaction between the three-phase sequences did not exist.

It is interesting to note that when, in the case of balanced loads,

$$Z_a = Z_b = Z_c = Z$$

from eqns. (231)

$$\zeta_0 = Z,$$

$$\zeta_1 = 0,$$

$$\zeta_2 = 0.$$

Substituting these values in eqn. (232) gives

$$\left.\begin{aligned} V_{a0} &= ZI_{a0}, \\ V_{a1} &= ZI_{a1}, \\ V_{a2} &= ZI_{a2}, \end{aligned}\right\} \tag{233}$$

These results are seen to confirm the previous theory for balanced loads, in which case there is no interaction between currents and voltages of different phase sequences.

66. Currents in unbalanced load

Further reference to Fig. 75 shows that, since there is no neutral current, the sum of the phase currents will be zero and so the zero-phase sequence component of current must also be zero, i.e.

$$I_{a0} = \tfrac{1}{3}(I_a + I_b + I_c) = 0. \tag{234}$$

Hence eqns. (232) may now be rewritten as

$$\left.\begin{aligned} V_{a0} &= \zeta_2 I_{a1} + \zeta_1 I_{a2}, \\ V_{a1} &= \zeta_0 I_{a1} + \zeta_2 I_{a2}, \\ V_{a2} &= \zeta_a I_{a1} + \zeta_0 I_{a2}. \end{aligned}\right\} \tag{235}$$

Since the applied voltages are balanced, and are assumed to consist of positive phase sequence components only, E_{a0} and E_{a2} will both be zero. Applying Kirchhoff's second law to the circuit for the positive- and negative-phase sequence voltages gives

$$V_{a1} = E_{a1} = E_a = \zeta_0 I_{a1} + \zeta_2 I_{a2}, \tag{236}$$

$$V_{a2} = E_{a2} = 0 = \zeta_1 I_{a1} + \zeta_0 I_{a2}, \tag{237}$$

which, from eqn. (237), leads to

$$I_{a2} = -\frac{\zeta_1}{\zeta_0} I_{a1}. \tag{238}$$

Substituting this result in eqn. (236) we have

$$E_a = \zeta_0 I_{a1} - \frac{\zeta_1 \zeta_2}{\zeta_0} I_{a1},$$

$$I_{a1} = \frac{E_a \zeta_0}{\zeta_0^2 - \zeta_1 \zeta_2}, \tag{239}$$

and also

$$I_{a2} = \frac{-E_a \zeta_1}{\zeta_0^2 - \zeta_1 \zeta_2}. \tag{240}$$

For the line currents we may now write

$$\left.\begin{aligned} I_a &= I_{a1} + I_{a2} = E_a \frac{\zeta_0 - \zeta_1}{\zeta_0^2 - \zeta_1 \zeta_2}, \\ I_b &= h^2 I_{a1} + h I_{a2} = E_a \frac{h^2 \zeta_0 - h \zeta_1}{\zeta_0^2 - \zeta_1 \zeta_2}, \\ I_c &= h I_{a1} + h^2 I_{a2} = E_a \frac{h \zeta_0 - h^2 \zeta_1}{\zeta_0^2 - \zeta_1 \zeta_2}. \end{aligned}\right\} \tag{241}$$

Alternatively, the positive- and negative-phase sequence components of current may be expressed in terms of the load impedances Z_a, Z_b, and Z_c where

$$I_{a1} = \frac{E_a\zeta_0}{\zeta_0^2 - \zeta_1\zeta_2},$$

$$= \frac{\frac{1}{3}E_a(Z_a + Z_b + Z_c)}{\frac{1}{9}(Z_a + Z_b + Z_c)^2 - \frac{1}{9}(Z_a + hZ_b + h^2Z_c)(Z_a + h^2Z_b + hZ_c)},$$

$$= \frac{E_a(Z_a + Z_b + Z_c)}{Z_aZ_b + Z_bZ_c + Z_cZ_a}, \tag{242}$$

and also

$$I_{a2} = \frac{-E_a\zeta_1}{\zeta_0^2 - \zeta_1\zeta_2} = \frac{-E_a(Z_a + hZ_b + h^2Z_c)}{Z_aZ_b + Z_bZ_c + Z_cZ_a}. \tag{243}$$

Similarly substitution of the load impedances in eqns. (241) gives the following results for the line currents:

$$\left.\begin{aligned} I_a &= \frac{(1 - h^2)E_a(Z_c - hZ_b)}{Z_aZ_b + Z_bZ_c + Z_cZ_a}, \\ I_b &= \frac{(1 - h^2)E_a(h^2Z_a - Z_c)}{Z_aZ_b + Z_bZ_c + Z_cZ_a}, \\ I_c &= \frac{(1 - h^2)E_a(hZ_b - h^2Z_a)}{Z_aZ_b + Z_bZ_c + Z_cZ_a}. \end{aligned}\right\} \tag{244}$$

67. Voltages across the unbalanced load

Considering now the voltages, we have, from eqns. (235) for the zero-phase sequence voltage,

$$V_{a0} = \zeta_2 I_{a1} + \zeta_1 I_{a2},$$

and substituting for I_{a1} and I_{a2} from eqns. (239) and (240),

$$V_{a0} = \frac{E_a \zeta_0 \zeta_2}{\zeta_0^2 - \zeta_1 \zeta_2} - \frac{E_a \zeta_1^2}{\zeta_0^2 - \zeta_1 \zeta_2},$$

$$= E_a \left[\frac{\zeta_0 \zeta_2 - \zeta_1^2}{\zeta_0^2 - \zeta_1 \zeta_2} \right]. \tag{245}$$

Expressing, now, the phase voltages across the load in terms of the phase sequence components gives

$$\left.\begin{aligned} V_a &= V_{a0} + V_{a1} = E_a \left[1 + \frac{\zeta_0 \zeta_2 - \zeta_1^2}{\zeta_0^2 - \zeta_1 \zeta_2} \right], \\ V_b &= V_{a0} + h^2 V_{a1} = E_a \left[h^2 + \frac{\zeta_0 \zeta_2 - \zeta_1^2}{\zeta_0^2 - \zeta_1 \zeta_2} \right], \\ V_c &= V_{a0} + h V_{a1} = E_a \left[h + \frac{\zeta_0 \zeta_2 - \zeta_1^2}{\zeta_0^2 - \zeta_1 \zeta_2} \right]. \end{aligned}\right\} \tag{246}$$

As in the case of the currents, the above voltages may also be expressed in terms of the load impedances Z_a, Z_b, and Z_c.

$$V_{a0} =$$

$$E_a \left[\frac{\frac{1}{9}(Z_a + Z_b + Z_c)(Z_a + h^2 Z_b + h Z_c) - \frac{1}{9}(Z_a + h Z_b + h^2 Z_c)^2}{\frac{1}{9}(Z_a + Z_b + Z_c)^2 - \frac{1}{9}(Z_a + h Z_b + h^2 Z_c)(Z_a + h^2 Z_b + h Z_c)} \right]$$

$$= - E_a \left[\frac{Z_b Z_c + h Z_a Z_b + h^2 Z_c Z_a}{Z_a Z_b + Z_b Z_c + Z_c Z_a} \right].$$

Alternatively, expressing the above equation in terms of admittances instead of impedances,

$$V_{a0} = - E_a \left[\frac{Y_a + h^2 Y_b + h Y_c}{Y_a + Y_b + Y_c} \right]. \tag{247}$$

The voltage of point p with respect to point q in the diagram will be equal to the zero-phase sequence voltage V_{a0}.

Inserting the values of the load impedances in eqns. (246),

$$V_a = E_a\left[1 - \frac{Z_bZ_c + hZ_aZ_b + h^2Z_cZ_a}{Z_aZ_b + Z_bZ_c + Z_cZ_a}\right],$$

$$= E_a\left[\frac{Z_cZ_a(1-h^2) - hZ_aZ_b}{Z_aZ_b + Z_bZ_c + Z_cZ_a}\right],$$

$$= (1-h^2)E_a\left[\frac{Z_cZ_a - hZ_aZ_b}{Z_aZ_b + Z_bZ_c + Z_cZ_a}\right],$$

or, using admittance terms instead of impedances,

$$V_a = (1-h^2)E_a\left[\frac{Y_b - hY_c}{Y_a + Y_b + Y_c}\right].$$

Similarly

$$\left.\begin{aligned} V_b &= (1-h^2)E_a\left[\frac{h^2Y_c - Y_a}{Y_a + Y_b + Y_c}\right], \\ V_c &= (1-h^2)E_a\left[\frac{hY_a - h^2Y_b}{Y_a + Y_b + Y_c}\right]. \end{aligned}\right\} \quad (248)$$

68. Use of unbalanced load to determine phase sequence

Two lamps and a capacitor, each of approximately the same ohmic impedance, may be used, when connected in star, to determine the phase sequence of a balanced three-phase supply. On being connected to such a supply, one lamp will be observed to burn more brightly than the other, indicating that the phase to which it is connected leads the phase to which the dim lamp is connected by 120°.

If, in Fig. 76, the lamps are represented by two equal pure resistances R connected in phases a and b, and the capacitor C is connected in the third phase, then

$$Y_a = Y_b = Y,$$

$$Y_c = jY,$$

since each branch is of the same ohmic value.

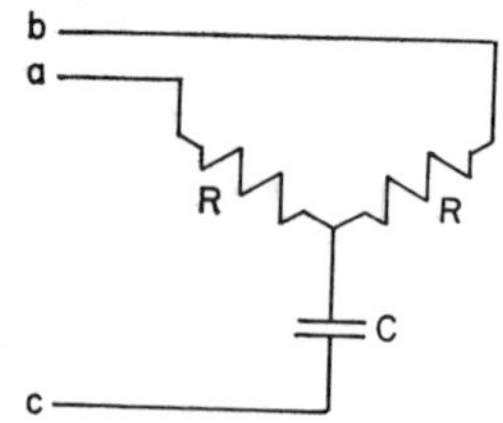

FIG. 76. Measurement of phase sequence.

Let the star-connected network be supplied from a balanced three-phase supply of phase voltage E and of phase sequence a–b–c. Using eqns. (248) the voltage V_a across the lamp connected to phase a is given by

$$V_a = (1 - h^2)E_a\left[\frac{Y_b - hY_c}{Y_a + Y_b + Y_c}\right],$$

$$= (\tfrac{3}{2} + j\tfrac{1}{2}\sqrt{3})E_a\left[\frac{Y - (-\tfrac{1}{2} + j\tfrac{1}{2}\sqrt{3})jY}{2Y + jY}\right],$$

$$= (\tfrac{3}{2} + j\tfrac{1}{2}\sqrt{3})E_a\left[\frac{1 + \tfrac{1}{2}\sqrt{(3)} + j\tfrac{1}{2}}{2 + j}\right],$$

$$= \frac{E_a}{5}(\tfrac{3}{2} + j\tfrac{1}{2}\sqrt{3})\,(1 + \tfrac{1}{2}\sqrt{(3)} + j\tfrac{1}{2})\,(2 - j),$$

$$= \frac{E_a}{5}(\tfrac{3}{2} + j\tfrac{1}{2}\sqrt{3})\,(\tfrac{5}{2} + \sqrt{(3)} - j\tfrac{1}{2}\sqrt{3}),$$

$$= \frac{E_a}{5}(\tfrac{18}{4} + \tfrac{3}{2}\sqrt{(3)} + j(\tfrac{3}{2} + \tfrac{1}{2}\sqrt{3})),$$

$$= E_a(1{\cdot}42 + j0{\cdot}473).$$

The magnitude of this voltage is then

$$|\, V_a \,| = E_a\sqrt{(1{\cdot}42^2 + 0{\cdot}473^2)},$$

$$= 1{\cdot}495\ E_a.$$

The voltage across the lamp connected to the b phase is given by

$$V_b = (1 - h^2)E_a\left[\frac{h^2 Y_c - Y_a}{Y_a + Y_b + Y_c}\right],$$

$$= (\tfrac{3}{2} + j\tfrac{1}{2}\sqrt{3})E_a\left[\frac{(-\tfrac{1}{2} - j\tfrac{1}{2}\sqrt{3})jY - Y}{2Y + jY}\right],$$

$$= (\tfrac{3}{2} + j\tfrac{1}{2}\sqrt{3})E_a\left[\frac{-1 + \tfrac{1}{2}\sqrt{(3)} - j\tfrac{1}{2}}{2 + j}\right],$$

$$= \frac{E_a}{5}(\tfrac{3}{2} + j\tfrac{1}{2}\sqrt{3})(-1 + \tfrac{1}{2}\sqrt{(3)} - j\tfrac{1}{2})(2 - j),$$

$$= \frac{E_a}{5}(\tfrac{3}{2} + j\tfrac{1}{2}\sqrt{3})(-\tfrac{5}{2} + \sqrt{(3)} - j\tfrac{1}{2}\sqrt{3}),$$

$$= \frac{E_a}{5}(-3 + \tfrac{3}{2}\sqrt{(3)} + j(\tfrac{3}{2} - 2\sqrt{3})),$$

$$= E_a(-0\cdot 08 - j0\cdot 393).$$

from which

$$|V_b| = E_a\sqrt{(0\cdot 08^2 + 0\cdot 393^2)},$$

$$= 0\cdot 401\, E_a.$$

The lamp connected to phase a will thus have a greater illumination than the lamp connected to phase b. If the phase sequence is reversed, the lamp in phase b will become the bright one, and hence such an unbalanced network may be used in the determination of the phase sequence of a balanced three phase supply.

Exercise

1. Two 60 W lamps and a pure inductance are connected in star. The ohmic value of the reactance of the inductance is equal to the resistance of one of the lamps. Show that, if the lamps are connected to phases b and c and the inductance to phase a of a balanced three-phase supply, the lamp connected to phase c will have a greater illumination than the one connected to phase b.

APPENDIX

Use of Matrix Notation

For the reader familiar with the use of matrix notation, some simplification in the manipulation of equations of symmetrical components may be obtained. Writing the expressions for the phase values of voltage in terms of the phase sequence components of the *a* phase we have

$$\left.\begin{aligned} V_a &= V_{a0} + V_{a1} + V_{a2}, \\ V_b &= V_{a0} + h^2 V_{a1} + h V_{a2}, \\ V_c &= V_{a0} + h V_{a1} + h^2 V_{a2}. \end{aligned}\right\}$$

Written in matrix form the above equations appear as

$$\begin{bmatrix} V_a \\ V_b \\ V_c \end{bmatrix} = \begin{bmatrix} 1 & 1 & 1 \\ 1 & h^2 & h \\ 1 & h & h^2 \end{bmatrix} \begin{bmatrix} V_{a0} \\ V_{a1} \\ V_{a2} \end{bmatrix} \tag{249}$$

which may be abbreviated to

$$[V] = [H]\,[V_s], \tag{250}$$

where $[V]$ $[H]$ and $[V_s]$ represent the matrices given by

$$[V] = \begin{bmatrix} V_a \\ V_b \\ V_c \end{bmatrix}$$

$$[H] = \begin{bmatrix} 1 & 1 & 1 \\ 1 & h^2 & h \\ 1 & h & h^2 \end{bmatrix}$$

$$[V_s] = \begin{bmatrix} V_{a0} \\ V_{a1} \\ V_{a2} \end{bmatrix}$$

In a similar way to eqn. (250) the phase values of current may be represented in terms of the symmetrical components by

$$[I] = [H]\,[I_s]. \tag{251}$$

It is also necessary to be able to write a matrix equation for the symmetrical components in terms of the phase values, in which case we write

$$[V_s] = [H]^{-1}\,[V],$$

where $[H]^{-1}$ is the inverse of $[H]$.

In order to find $[H]^{-1}$ it is first necessary to find the determinant

$$|\,H\,| = \begin{vmatrix} 1 & 1 & 1 \\ 1 & h^2 & h \\ 1 & h & h^2 \end{vmatrix}$$

$$= (h^4 + h + h) - (h^2 + h^2 + h^2),$$

$$= 3(h - h^2).$$

The transpose of $[H]$ is then found where

$$[H]_t = \begin{bmatrix} 1 & 1 & 1 \\ 1 & h^2 & h \\ 1 & h & h^2 \end{bmatrix}$$

The inverse of $[H]$ is now given by

$$[H]^{-1} = \frac{1}{3(h-h^2)} \begin{bmatrix} h-h^2 & -(h^2-h) & h-h^2 \\ -(h^2-h) & h^2-1 & -(h-1) \\ h-h^2 & -(h-1) & h^2-1 \end{bmatrix}$$

$$= \frac{1}{3} \begin{bmatrix} 1 & 1 & 1 \\ 1 & h & h^2 \\ 1 & h^2 & h \end{bmatrix}$$

Hence we may now write

$$[V_s] = [H]^{-1}[V]$$

$$\begin{bmatrix} V_{a0} \\ V_{a1} \\ V_{a2} \end{bmatrix} = \frac{1}{3} \begin{bmatrix} 1 & 1 & 1 \\ 1 & h & h^2 \\ 1 & h^2 & h \end{bmatrix} \begin{bmatrix} V_a \\ V_b \\ V_c \end{bmatrix}$$

thus giving the already established equations

$$\left.\begin{aligned} V_{a0} &= \tfrac{1}{3}(V_a + V_b + V_c), \\ V_{a1} &= \tfrac{1}{3}(V_a + hV_b + h^2V_c), \\ V_{a2} &= \tfrac{1}{3}(V_a + h^2V_b + hV_c). \end{aligned}\right\}$$

Bibliography

CLARKE, *Circuit Analysis of A.C. Power Systems* (Wiley).
DAHL, *Electric Circuits—Theory and Applications* (McGraw-Hill).
FORTESCUE, Method of symmetrical coordinates applied to the solution of polyphase networks, *Trans. A.I.E.E.*, Vol. 37, Part II.
LACKEY, *Fault Calculations* (Oliver & Boyd).
LYON, *Applications of the Method of Symmetrical Components* (McGraw-Hill).
MORTLOCK and DAVIES, *Power System Analysis* (Chapman & Hall).
RISSICK, *The Calculation of Unsymmetrical Short Circuits* (Pitman).
WAGNER and EVANS, *Symmetrical Components* (McGraw-Hill).

Answers to Exercises

CHAPTER 1

1. $\frac{3}{2} - j\frac{1}{2}\sqrt{3}$; $\frac{1}{2} + j\frac{1}{6}\sqrt{3}$; $\frac{1}{2} - j\frac{1}{2}\sqrt{3}$.
2. $(0 \cdot 38 + j1 \cdot 66)$; $(2 \cdot 86 + j0 \cdot 48)$; $(1 \cdot 66 + j0 \cdot 38)\ \Omega$.
3. 2 p.u.; 1 p.u.

CHAPTER 2

2. $58\underline{/43 \cdot 2°}$; $19\underline{/25°}$; $27 \cdot 3\underline{/4 \cdot 7°}$; 81·9 A.
3. 0·51.
4. $1938\underline{/9 \cdot 9°}$ A; $261\underline{/9 \cdot 7°}$ A; $915\underline{/-24 \cdot 4°}$ A.

CHAPTER 3

1. 3 kW; 0.
2. 3190 A; 581 A.

CHAPTER 4

1. 3; 1·5 V.
2. Line voltage; zero.
3. $17 \cdot 4\underline{/150°}$ V.

CHAPTER 5

1. 138 A; 16·9 kW.
3. $152 \cdot 5\underline{/-10 \cdot 9°}$ A; $57 \cdot 5\underline{/-90°}$ A; 573 V.

CHAPTER 7

1. 5445 A; 276 A; 5445 V.
2. 5500 A; 12,720 V.

Index